Samuel F. B. Morse and the Dawn of the Age of Electricity

Samuel F. B. Morse and the Dawn of the Age of Electricity

George F. Botjer

LEXINGTON BOOKS
Lanham • Boulder • New York • London

Published by Lexington Books
An imprint of The Rowman & Littlefield Publishing Group, Inc.
4501 Forbes Boulevard, Suite 200, Lanham, Maryland 20706
www.rowman.com

Unit A, Whitacre Mews, 26-34 Stannary Street, London SE11 4AB

Copyright © 2015 by Lexington Books

All rights reserved. No part of this book may be reproduced in any form or by any electronic or mechanical means, including information storage and retrieval systems, without written permission from the publisher, except by a reviewer who may quote passages in a review.

British Library Cataloguing in Publication Information Available

Library of Congress Cataloging-in-Publication Data

Botjer, George F., 1937-
Samuel F.B. Morse and the dawn of the age of electricity / George F. Botjer.
pages cm
Includes bibliographical references and index.
ISBN 978-1-4985-0140-8 (cloth : alk. paper) —ISBN 978-1-4985-0141-5 (electronic)
1. Morse, Samuel Finley Breese, 1791-1872. 2. Inventors—United States—Biography. 3. Painters—United States--Biography. 4. Telegraph—History. I. Title.
TK5243.M7B68 2015
621.383092—dc23 [B]
2015004059

∞ ™ The paper used in this publication meets the minimum requirements of American National Standard for Information Sciences Permanence of Paper for Printed Library Materials, ANSI/NISO Z39.48-1992.

Printed in the United States of America

This book is dedicated to my parents

Contents

Preface		ix
A Note on Sources and Endnote Citations		xi
Bibliographical Note		xiii
1	An American Artist: Fame and Misfortune	1
2	Starving Artist Invents Telegraph in Greenwich Village Garret	21
3	From Wilderness to Empire: Morse and the System Builders	39
4	The Question of Origins and Originality: Did Morse Really Invent the Telegraph?	67
5	The Great Man Revered and Reviled	93
6	Locust Grove	109
Concluding Remarks		123
Bibliography		125
Index		129
About the Author		133

Preface

My interest in Samuel F. B. Morse and the Morse Telegraph had its origins in a popular course that I taught for a number of years, as part of a capstone program for college seniors. Titled, "Technology and Social Change in America," it sought to combine three related fields of scholarship—economic history, economic geography, and the history of technology—into a seamless narrative covering the entire span of American history. The issues of timeliness ("right time") and location ("right place") were prominent features throughout, along with close-up studies of the great inventors.

In the select group of breakthrough inventions, the Morse Telegraph initially didn't stand out in comparison with, say, Fulton's steamboat and Edison's light bulb. Yet over time, through successive course offerings, the telegraph began to loom ever larger in the impressive and rather unique panorama of American inventiveness.

This change in valuation was based, to a large extent, on a greater attentiveness to the difference between "exponential" and "incremental" inventions. The telegraph belonged to the former category, for the simple reason that nothing at all like it had ever existed before. (The telephone, by comparison, was incremental—its parent being the telegraph.) Then, having established that distinction, there was the issue of social impact—a central concern of the course. Among the other exponential breakthroughs (e.g., the atomic bomb), where did the Morse Telegraph stand in terms of social impact? I eventually determined that it stood at or near the top of the list. It was, in short, one of the most original and important inventions of all time.

That conclusion made the question of why Morse (and nobody else) invented it all the more intriguing. This book is the result of my search for the answer to that question.

The latter part of the book (chapters 5 and 6) takes up a related matter that was also a part of "Technology and Social Change in America": What does a great invention do to and for its inventor(s)? And what does the answer to that question say about the society that nurtured those inventors?

Thanks are due to the University of Tampa for its encouragement of my research and writing, via a sabbatical and summer research grants. Thanks are due especially to my colleagues on the University's faculty development committee. Thanks also to the librarians at the Library of Congress, the Smithsonian Institution, and the New-York Historical Society. Thanks also to Kay Peterson, at the Smithsonian; Janet Bunde, acting archivist at New York University; Brenda B. Bradley, art historian and technical adviser; and Laura Gicker, who assisted in preparing the manuscript.

A Note on Sources and Endnote Citations

All of the M (Morse) citations in the endnotes refer to letters in the General Correspondence and Related Documents or the Letter Books in the Samuel F. B. Morse Papers (Library of Congress, Manuscript Division). The General Correspondence and the Letter Books are cataloged chronologically in folders, and within each folder, the letters are arranged chronologically. Except for the rare instances of undated material (for which there are folders in the General Correspondence), the endnotes include the date of each letter. The General Correspondence covers Morse's entire lifetime. In 1845, however, the inventor began filing letters he wrote in separate folders called Letter Books, while the incoming mail continued to be saved in the General Correspondence. The endnotes nonetheless treat the entire correspondence as a unitary Letter File. Endnote 29 in chapter 1, for instance, reads as follows: Cooper–M, 17 November, 1849; M–Cooper, 25 November 1849. The complete incoming letter (from James Fenimore Cooper) can be found in the appropriately dated folder in the General Correspondence. The reply from M (Morse), however, is in the 1849 Letter Book.

Bibliographical Note

Most of the material used in this book comes from the vast collection of Samuel F. B. Morse Papers housed in the Manuscript Division of the Library of Congress. Also consulted were the Alfred Vail Telegraph Collection at the Smithsonian Institution, the Henry O'Reilly (O'Rielly) Papers, at the New-York Historical Society, and—all at the New York Public Library—the Francis O. J. Smith Correspondence, the Western Union Archive, and the Cyrus Field Papers.

The three prior general biographies of Morse—Samuel I. Prime, *Life of Samuel F. B. Morse* (NY: Appleton, 1875); Carleton Mabee, *The American Leonardo* (New York: Knopf, 1943); and Kenneth Silverman, *Electric Man* (New York: Knopf, 2003), were also consulted, along with the three biographies by art historians—Oliver W. Larkin, *Samuel F. B. Morse and American Democratic Art* (Boston: Little, Brown 1964), William Kloss, *Samuel F. B. Morse* (New York: Abrams 1988), Paul J. Staiti, *Samuel F. B. Morse* (Cambridge: Harvard University Press, 1989). These latter works surpass the general biographies in their level of scholarship, but they deal only with Morse the artist.

General histories of the telegraph provided valuable insights and background information. A work by Morse's associate, Alfred Vail, *The American Electro-Magnetic Telegraph* (Philadelphia.: Lea & Blanchard, 1845), was indispensable. So too was the work of James D. Reid, *The Telegraph in America: Its Founders, Promoters and Noted Men* (New York: Derby Brothers, 1879). As one of the telegraph pioneers himself, Reid provides an excellent insider's perspective. The published papers of Joseph Henry (Washington, D.C.: Smithsonian Institution Press, 1972–2008), Volumes 7 and 8, contain material on his dispute with Morse. Good recent histories include David Hochfelder, *The Telegraph in America, 1832–1920* (Baltimore: Johns Hopkins University Press, 2012); Menahem Blondheim, *News Over the Wires: the Telegraph and the Flow of Public Information in America, 1844–1897* (Harvard University Press 1994); Lewis Coe, *The Telegraph: a History of Morse's Invention and its Predecessors in the United States* (Jefferson, NC: McFarland 1997); and Brian Winston, *Media Technology and Society: a History from the Telegraph to the Internet* (New York: Taylor and Francis, 1998). A book by George B. Prescott, *History, Theory and Practice of the Electric Telegraph* (Cambridge: Welch, Bigelow & Co., 1863) describes, very briefly, Harrison G. Dyar's experiment at the Long Island racetrack. It is the longest comment on Dyar in the entire literature, although Coe's work also has a brief mention of Dyar. An article in *Technology and Culture* (Vol. 11, No. 4, October 1970, p 493–549) titled "Art, Technology and Science: their Historical Interaction," by Cyril Stanley Smith contains insights into the creative and integrative capabilities that an artist (such as Leonardo da Vinci, Robert Fulton, and Morse) can bring to science and technology.

Of Morse's important associates, only Cyrus Field has inspired any significant biographical attention. Among the works on Field, a biography by Samuel Carter, *Cyrus Field: Man of Two Worlds* (New York: Putnam 1968) was by far the most informative and insightful.

There is nothing comparable on Ezra Cornell, and no full-length biography of Francis O. J. Smith. Everything I found on the other pioneers of the industry came primarily from the Morse Papers, with some information from the general telegraph histories.

Chapter One

An American Artist

Fame and Misfortune

Of the seven Morse biographies, four were written by art historians who had no interest whatever in his subsequent career as an inventor. All ranked him as a major artist of the early National period in U.S. art history, and Morse paintings are prominently displayed today in major museums, including the Metropolitan in New York. The Met contains his last painting, which was completed in 1843, in the same NYU studio where he invented the telegraph. It is a full-length portrait of his oldest child, Susan. Morse was at the height of his powers as an artist at that moment, but by deliberate choice he would not touch brush to canvas again for the remaining twenty-nine years of his life.

Samuel Finley Breese Morse was also a founder and longtime president of the National Academy of Design, which became the center of Gotham's art life and the home of the Hudson River School of painters.

So, first an artist and then an inventor. There is much speculation about the "creative imagination" and how a talent for graphic design and composition can serve the cause of industrial design, where form necessarily follows function. Robert Fulton, the inventor of the steamboat, was an artist. Leonardo da Vinci, the great Renaissance painter, was an architect of fortifications. Visualization and a more abstract "vision" are surely important for inventors. They were essential to Morse the inventor.

Morse the artist started out very young. A story from his infancy has him marking up the walls and furniture of a babysitter with crayons and chalk. A diary entry when he was fourteen reads, "I looked over my leson [sic] and then began to draw."[1] Another says, "…painted a few pictures."[2] Similar entries competed with his thoughts on religion as the main topic of the diary.

He drew and painted all through his school years. A painting of "Freshmen Climbing the Hill of Knowledge," from his Yale years, now hangs in a Chicago museum. His first art lessons didn't take place, however, until after graduation in 1810. They were actually anatomy classes for medical students, in Boston. He found the experience "extremely interesting" and "not disgusting" as he had expected.[3]

After graduating from Yale, Morse found employment at the Farrand & Mallory Bookstore in Boston. Mr. Mallory was a friend of his father. A well-known artist named Washington Allston was an occasional customer, and Morse one day got to show him a painting he had just completed of the Pilgrims arriving at Plymouth Rock.

Allston was preparing for a trip to London, where he was going to teach drawing and painting at the British Academy. There were no American counterparts to that prestigious institution. Morse, indeed, had never met a professional artist before.

Impressed by the Pilgrims picture, Allston offered Morse a chance to study at the British Academy. He was confident that the young man could pass the entrance exam. The main problem was paying for two or three years abroad.

His parents, Jedidiah and Elizabeth Ann were well aware of their oldest son's lifelong interest in art. But they had taken care to provide him with a good education, in the hope that he would embark on a career in one of the learned professions. His most distinguished ancestor, E. Samuel Finley, was the longtime head master of Nassau Hall, Princeton College, and Mrs. Morse's great uncle. Mom always addressed her son as Finley, and for all his life, so did everybody else. (Breese was Elizabeth Ann's family name.)

Jedidiah was the pastor of the First Congregational Church of Charlestown, MA (just north of Boston), a post which he had held since 1789. The son of Connecticut farmers, and the first in his family ever to attend college, Jedidiah had a Doctor of Divinity degree from Yale. By the time his oldest son went to Yale, Jed was a well-known figure among churchgoers all over New England. He edited a widely distributed religious monthly, *The Panoplist*, and was a founding trustee of the Andover Theological Seminary. In addition, he was an officer of the American Bible Society, the Tract Society, and the American Board of Commissioners for Foreign Missions.

Jedidiah Morse led a separate life as the author of geography books—the first ever written by an American. His *American Universal Geography*, first published in 1793, and updated regularly ever since, was required reading in schools and colleges all over the country. He also produced an almanac titled *The American Gazetteer*, which was updated at irregular intervals.

Morse went to school at Phillips Academy, in Andover, MA. He displayed a notable lack of interest in academics, and the prospect of not going to college seemed very real. Jedidiah and Elizabeth Ann, however, were quite determined. He was admitted to Yale because his father was a friend of the college president, Dr. Dwight (who once spent a night at the Morse parsonage while visiting Boston). Dwight acceded to a conditional admission, with a light class schedule the first term. Morse pulled through this trial, and became a regular student in 1806. According to a diary he kept, the greatest challenge during that dodgy first term was the nightly homework assignment of translating 126 lines from Homer's Greek epics. He would burn the midnight oil while classmates, having completed their work for the day, were off somewhere enjoying themselves.

Morse stuck it out to the end, and graduated in 1810. Part of the motivation was provided by his two younger brothers, Richard and Sidney, who were respectively one and two years behind him. Both had also followed him through Phillips Academy, and both were excellent students. Sidney, indeed, made Phi Beta Kappa at Yale.

His parents made a commitment to support him for up to three years in England, with an allowance of $1,000 a year. Morse quit the bookstore (which, as it happens, was the only regular job he ever had in his entire life), and set out for Gotham with Mr. and Mrs. Allston, to catch a boat for England. The fare to Liverpool took the first $90 of his annual allowance, and a coach trip from Liverpool to London cost another $30.

The British Academy's annual exhibit of new (mainly student) art was a major cultural event in London. Daily instruction throughout the year was largely limited to drawing classes, however. Morning sessions with live models were its principal activity. The paintings and sculpture that made up the annual show were often done in the students' living quarters—rented rooms in the neighborhood of the Academy.

Morse found a furnished room near the Academy, and only a block from the Allston household. He shared it with a fellow American, Charles Leslie, and they sometimes took in a boarder to help with the rent. Morse did all of his painting there. One of his earliest endeavors in London was a rendition of "Marius in Prison," which was based on a picture in the Academy's collection. Morse reported to his parents that Allston visited occasionally to view his current work, and was always very critical but also very encouraging.[4]

He rarely saw the head of the Academy, the renowned Benjamin West— an American expatriate famous for his portraits of George Washington. But Allston reported that West had been impressed with the drawings the young man submitted for admission, and gave prominent display to the work he submitted for the annual exhibition.

When the War of 1812 broke out, none of the Americans at the Academy experienced any change in their daily routine. The only inconvenience was the routing of mail back home, which had to go through a neutral port, and was therefore slower than usual. West, the portrayer of George Washington, remained as director.

In 1812, Morse was painting "the dying Hercules," and had a two-foot-high clay statue of the writhing strongman as a model. Allston was very impressed with the statue, and told Morse it was the best thing he had ever done. Bringing Benjamin West to the student's humble room, Allston exclaimed, "I have always told you any painter can make a sculpture."[5] The young artist went ahead with the painting.

In May 1813 both painting and statue were prominently displayed at the Academy's annual exhibit. The sculpted Hercules was awarded a Gold Medal as the best sculpture of a single figure. The medal was presented to the twenty-two-year-old American by the earl of Cumberland, a member of the royal family, before an audience of lords and ladies and other prosperous Londoners.

Morse's art historian biographers have noted the exact resemblance of the Hercules to a statue and painting by the Italian Renaissance artist Carlo Reni. William Kloss adds that many Renaissance artists made clay figures of their main subjects, to use as models, before touching brush to canvas. Kloss says, further, that much painting in Morse's day was a "rhetorical amalgam of borrowings from antiquity and the Old Masters."[6] Morse's painting was distinctive enough, in its coloration, its slightly different vantage point (more frontal and close-up than Reni's), and in the positioning of Hercules's legs.[7]

The prize-winning statue is different from the Reni painting only in the positioning of the legs. The contorted musculature and posture of the subject was about as difficult a challenge for a sculptor as one could imagine. It was not the sort of thing a beginner in sculpture would have even imagined trying. But the art historians have Allston's imprimatur, and that apparently is enough for them.

Morse sent a plaster cast of the clay model of Hercules home to Charlestown, along with the painting.[8] In 1860, somebody asked him where they were. The former artist replied that he had given the statue to a Reverend E. G. Smith, in Washington, "years ago." The six-by-eight-foot painting had been donated to the Yale Art Gallery, after Morse bought a $100 frame for it. A plaster cast of the statue was also given to Yale.[9]

In 1815, the artist asked his parents to add another year to their three-year commitment, and support him for a year. He said most Academy students completed their training on the Continent, adding that art, among all the "professions," was the only one that required travel to see what colleagues in other countries were doing.[10]

His father, Jedidiah, had previously advised the young artist that the $1,000 a year in London was straining his resources. He was still supporting Morse's two brothers, Richard and Sidney. Both had by now graduated from Yale and were attending Andover Theological Seminary. Their upkeep combined was less than Morse's. Since Jed's salary at the church amounted to only $1200 a year, all this would have been unaffordable without the steady drizzle of royalties from the famous geography textbook, *American Universal Geography*. It still had no serious competition in the classrooms of America. But Jed's finances had taken a blow when he invested $5,000 and in a private toll road project in New York State. He had recently taken several weeks' leave from the church in order to go hunting for the people who had separated him from his nest egg, but to no avail.

Nonetheless, he agreed to a fourth year, but only if his son remained in London. Jed's concern here was to keep the young man, who had proven to be a spendthrift in his early days abroad, under the close supervision of a local bank officer he had hired to manage Morse's finances.

The artist found this condition unacceptable, and concluded sadly that his foreign studies were at an end. He arrived home in October 1815, and then began wandering all over New England in search of paying sitters. His portraits went for $15 to $85. The prospect of a group portrait of the Dartmouth faculty, for $500, fell through when several professors decided not to kick in any money. Some of his subjects were important people—mayors and judges, and also Gov. Langdon of New Hampshire, whose portrait was done for $100, and placed on display at the State House.

Morse first met his future wife, Lucretia, during a brief sojourn in Concord, NH, in 1816. She was just sixteen years old. Thin as a rail, very quiet and in delicate health, she was as different from the robust, buxom, and gregarious Mrs. Morse as anybody could be. They became engaged in 1817, after the young girl spent some time as a guest at the Charlestown parsonage. Morse was "on the road" during that entire visit, and Mr. and Mrs. Walker, her parents, entertained serious doubts about the artist's financial stability and obviously bohemian proclivities. In January 1818, Morse decided to spend a "season" (winter) in Charleston, SC. His uncle, Dr. James Finley, was a prominent physician there, and (in reply to an inquiry from the artist) said there were plenty of people in town with money to spend who were eager to have their portraits painted.

With the onset of sultry weather in late April, Morse headed back north. He had several unfinished commissions to work on, and $3,000 in his pocket. This was very reassuring to the Walkers. The artist married Lucretia Walker in August, and a month later the two of them booked passage on a commercial schooner out of Boston, bound for Charleston. The Morses stayed briefly at the Finley house on King St., near St. Michael's, the church with Old Charleston's iconic steeple. They soon found a cottage for rent nearby, and the artist set up his portrait studio there.

Morse completed twenty-seven commissions during that second season in Charleston. He and Lucretia returned to New England in March 1819, with forty unfinished portraits in their luggage.

In the fall of 1819, Morse returned to Charleston, leaving Lucretia with his parents in Charlestown. She had been ill much of the time during their

Carolina "honeymoon" the year before, and both suspected the Charleston climate. The major highlight of this third season was a $500 commission from the City to paint a portrait of President James Monroe when he visited in December. Morse ended up having to follow the president back to Washington, where a couple of weeks of waiting yielded two more sittings with the president. He was rewarded with an invitation to the presidential New Year's Day reception, and, a few nights later, dinner with the First Family at the White House.[11]

The season, overall, showed signs of exhausting the market. Several other painters had shown up to compete with Morse. The Monroe and two large commissions from local plantation owners, who wanted paintings of several family members, made the season worthwhile. But such were not likely to be repeated. In March, the artist wrote his parents that his hopes for enough financial independence to devote all his time to "large…historical pictures" were now in question. By season's end, as the heat and the mosquitoes began to take hold, Morse had cleared $4,000.[12]

He returned to Charleston for a fourth time in November 1820—later than usual, and fairly certain that this would be his last season there. A wealthy planter named William Alston (no relative of Washington Allston) commissioned multiple portraits of every member of his large family that year. He made detailed requests for elaborate backgrounds that would show off rooms in his Georgian-style mansion, or, occasionally, the grounds of his plantation, which was home to one hundred slaves. Morse sometimes spent several days at a time there, with Alston frequently looking over his shoulder and making suggestions.

The artist had one difficult client that season, a Mrs. Caroline Ball, who complained (after making a down payment) that his $800 price was "exorbitant." The artist replied that his just-completed portrait of Mrs. Alston had cost the same amount, and that Mr. Alston was so happy with it that he gave the artist an extra $200. Mrs. Ball then paid the $400 balance due with a check, and took the picture away. The check bounced.[13]

Most of the artist's portraits were small and simple head-and-torso compositions that he sold for $50 to $80. He cleared $3,200 in this final season, on sixty commissions. More than a third of the money came from Mr. Alston, who had commissioned at least a dozen paintings altogether. A newspaper ad in March 1821 advised that the artist was about to leave for good, and invited last-minute commissions. A flurry of new work resulted, causing him to remain until April.[14]

After Charleston, the artist found himself at loose ends. One highlight of this period was a commission to paint the U.S. House of Representatives, which began in November 1821. He was assigned a small room right next to the House Chamber to use as a studio. Members proved quite willing to stop by, in accordance with a posted appointment schedule, and sit briefly for a likeness that was sketched in pencil. Several members, impressed with the drawing, paid $100 each for a painted portrait, and came back for another sitting or two.

Morse completed some of these before leaving Washington. The remainder was done in New Haven, CT, where he was living in his parents' rented house. The large picture of the entire House was painted there, too. The artist even paid to have a new room added to the house, to serve as a studio.

When the panoramic scene of the entire House was completed, in 1822, the artist was confident that he had captured, in each tiny face, a recognizable likeness. It was an authentic large-group portrait, a rarity in that pre-photographic era. Oddly enough, though, it was not sent back to Washington for examination by its subjects. Instead, it was shipped to New York, in January 1823, and put on exhibit for paying customers.

In February, after a not very rewarding sojourn in Gotham, the large canvas went to Boston, where Allston saw it. It then travelled around New England, initially under the care of an art dealer named Henry C. Pratt, and later his associate, Thomas Doolittle. Admission fees, at least as reported by Pratt and Doolittle, barely paid rental and transportation expenses, and the artist didn't make a dime. There is no evidence that he tried to sell it to the House of Representatives, although one must assume that an attempt was made at some point. The fact that the 1822 election saw about 20 percent of the "sitters" replaced by people who were not in the picture was certainly not a "plus."

The young artist at this point had barely $150 in the bank, a wife, and two children to support, and no job.[15]

The painting ended up in a storage shed, alongside the large Hercules canvas that had been sent home from England. (In more recent times, it found a permanent home at the Corcoran Gallery, in Washington.)

Morse found himself back at square one of his art career: hitchhiking around New England, looking for sitters. August 1823, for instance, found him in Albany, NY, where he put an ad in the local newspaper and waited at his hotel for an inquiry. None came, and he paid his room rent with a portrait of the manager and his daughter.

Where was his wife? When the newlyweds returned from Charleston early in 1819, Lucretia settled in at the parsonage. Later that year, she gave birth to a daughter, Susan.

Not having a home of her own was disappointing enough. The parsonage, which parishioners had built for Jed and his wife in 1791 (the year Morse was born) was anything but a nest of tranquility at this time. Jed was in financial straits, and also having trouble with his parishioners. The two problems were indirectly connected. The pastor had co-signed a loan for his longtime friend, Mr. Mallory, of the Farrand & Mallory Bookstore in Boston. When Mallory went broke, the entire debt fell on him like a ton of bricks. Jed lost all of his savings. He then decided on a hurried update of his almanac, and at the same time, a new edition of the best-selling geography book.

These endeavors proved a distraction from his duties at the First Congregational Church. Members began to grumble about a lack of home visits, and a scarcity of guest minister exchanges with other churches in the region. Attendance fell off, and some memberships were not renewed. Then, in 1820, the thirty-first year of his ministry at the church, a faction petitioned for Jed's resignation. A panel of fellow ministers was brought in to investigate the alleged problems, and mediate a possible solution. When it failed to resolve the issues, Jed resigned. The family, including Lucretia and her infant daughter, then moved to New Haven, where they rented a house.

Through all of this, the family was so beset by bill collectors that Jed resorted to selling his copyright in the hope of obtaining quick cash. He sold the rights for individual states, but usually, instead of hard cash, he settled for short-term promissory notes. Notes were defaulted, and royalties went into the wrong pockets, pending court proceedings, which Jed could ill afford. Bill collectors came knocking with regularity, and would hang around in the front yard when nobody answered the door.

In 1821 Lucretia had a miscarriage, and another followed in 1822. Then, in 1823, she gave birth to a son, named Charles. Her last child, James, was born in 1824. Morse, during this period, was gone for two or three months at a time. He was not present for any of these birth events.

In 1824 the wandering artist took a boat down the Hudson to Gotham, and rented a tiny unfurnished room on lower Broadway. The room had a small wooden table and chair, but no bed. He slept on the floor.

His two brothers, Richard and Sidney, had moved to New York the previous year and founded a religious weekly named the *New York Observer*. Richard sometimes delivered papers to local churches by pushcart. On arrival, Morse asked his two brothers for a small loan, but neither had any money to spare.[16]

Morse became a habitué of Lyman Trumbull's American Academy of Fine Art, which was a short walk from his rented room in the Wall Street area. There he met a number of artists and students. Many of them had heard of him. He was by far the best trained artist in the group, a bit older than most, and the only one with a college degree. The fact that he had painted not only the U.S. House of Representatives but also President James Monroe, John Adams (in 1816) and was soon commissioned to paint Eli Whitney (in 1824) lent him still greater status among the students at the American Academy.

Trumbull was quite well known at that time. Reproductions of his Revolutionary War paintings could be found in every history textbook in every schoolhouse in the land. (And they still turn up in many modern texts.) He ran the school, however, as an autocrat. The trustees—local business and professional people who provided modest support—gave him free rein. The Academy was actually just a large room with several plaster casts that were used as sketching models, and a small library with about two dozen art books. There were no scheduled classes, no instructors, and no regular hours.

The lack of regular hours provided the spark that ignited a full-scale rebellion, one cold morning in January 1826. Students hoping to light up the stove inside and make themselves warm (after a cold night, for most of them, in their various hovels) found the door locked. There was no telling if Trumbull would show up at all that day and let them in. Thirty of them—Morse included—signed a statement announcing their withdrawal from the school, and stuck it in the door. They then found refuge in a nearby coffee house, where they decided to set up their own art school. It was somewhat grandiosely named the National Academy of the Arts of Design.[17]

Morse was a ringleader of the secession—or so Trumbull was later told. He went to the City Clerk and obtained a corporate charter for the new institution. The students then elected him as the (unpaid) "Director" of their new institution.

Several of the trustees of Trumbull's school switched their allegiance to the new Academy. Morse played an active role in soliciting this support, and in fact, he secured some commissions while he was at it. The down payments allowed him to move out of his cold and tiny garret and rent two furnished rooms with a fireplace in a physician's house a few blocks up Broadway. Meanwhile, the new Academy found a cheap commercial space nearby. The first of its annual exhibits (which have been continuously held ever since) took place in May 1826.

Morse was a virtual pauper when he first arrived in the Big City, but his background as an alumnus of the British Academy who had painted the portrait of James Monroe certainly contributed to his rapid rise as one of Gotham's prominent cultural figures. The new Academy owed much to his early leadership, just as he owed much to the Academy as a vehicle for his rise to cultural prominence in the Big City. With a board of trustees that included a former mayor and some of the city's most prominent citizens, Morse found himself hobnobbing with Livingstons, Rensselaers, Kents, et al. He also struck up a friendship with the famous young novelist James Fenimore Cooper at this time, and became acquainted with William Cullen Bryant, the noted poet and editor of the *New York Evening Post*.

These three—Morse, Bryant, and Cooper—formed the nucleus of an informal coffee house gathering that they called "the Lunch." Students from the Academy and aspiring writers who wanted to be around Cooper and Bryant became "regulars," and it was not uncommon for the politicians and members of the merchant and professional classes to drop by and rub shoulders with these cognoscenti. Bryant was the main attraction for the politicians, because of his position at the *Post*, which had the largest circulation of any Democratic Party-affiliated newspaper in the country.

In March and April 1826, Morse was invited to give four lectures on art at Columbia University. This was part of the university's Athenaeum Lectures series, which included science and philosophy as well as the arts. (He had, in fact, already attended Athenaeum lectures on electro-magnetism.) His presentations took place on four consecutive Saturday evenings, and dealt with the relationship of art and society—how society influences art, and vice versa. The notes he prepared comprise his only writings on art.[18]

The Marquis de Lafayette, France's liaison to George Washington during the Revolutionary War and a regimental commander during the climactic battle of Yorktown, arrived in the United States in January that same year for the country's fiftieth birthday. New York City commissioned Morse to paint his portrait. Rather than take his chances on at most one or two sittings when the distinguished visitor arrived in Gotham, the artist managed to arrange the first sessions while Lafayette was still in Washington. He took a stagecoach down to the Capital City.

The first sitting went well. But then a letter arrived from brother Richard, with the terrible news that Lucretia, had died suddenly at the rented house in New Haven. He dropped everything and boarded the next coach north, but was a day late for the funeral.

When artist and sitter met again weeks later, in New York, the Frenchman seemed genuinely concerned about this tragic turn of events. Morse was not eager to talk about his wife—the wife to whom he had given so little; least of all his time and attention; most of all—three little children with no visible means of support. Lafayette did not ask questions. Instead, during two long sessions, they fell into lengthy discussions about the future of democracy and the other great issues of the day. Both agreed that democracy was the wave of the future, and that the forces of reaction would not be easily overcome.

The large, full-length painting was completed in June, just in time for unveiling at a gala reception. The picture was to be a lasting memorial of this memorable visit, and was assigned a prominent place in the city hall rotunda.

The artist did portraits of some local judges, for $100 each; a commissioned history painting of "Prince Arthur" (for $150); a couple of $15 landscapes, a portrait of General William Ledyard, and paintings of various businessmen and officials and their wives and children. He also had a first taste of notoriety, when the *North American Review* (a Boston journal) ridiculed the "pretentious" name of his school, which was now called the National Academy of Design.[19]

Morse did not earn a salary as head of the Academy, but in November 1829, the trustees decided to send him to Europe, all expenses paid. Each subscribed $100 to $300, and each was promised a picture of some European subject, painted "from life" during the trip. Some requested copies of specific Renaissance masterworks. Others left the choice of subject to the artist.

Washington Allston, whose contacts with Morse had been sporadic since 1815, wrote from Boston asking him to take tracings of historic bas reliefs, and to write regular letters. Morse replied that he would do what he could. (Oddly enough, that turned out to be very little.)

His brothers relayed an offer from the *Journal of Commerce* to write a weekly newsletter while on his travels. Richard and Sidney knew that their brother could write well enough, and it was they who brought the idea to their colleagues, who worked just down the street from the Observer's offices near Wall Street. What especially inspired them to do this was the recent publication (earlier in 1829) of a book of poetry—not written by Morse, but collected and edited by him. Styling himself "Professor Morse," he related in an introduction how, while passing through Plattsburgh, NY, he had been shown a sheaf of poems by his hosts, a Mr. and Mrs. Davison. The poems were the work (presumably) of their recently deceased teen-aged daughter, Lucretia, and had been found in a closet. Struck by the elegance of the writing, Morse promised the Davisons that, if they would entrust their precious discovery to him, he would make sure that it was published, for the entire world to marvel at and enjoy.

The resulting very thin volume, "Amr Khan and Other Poems," failed to take the poetry-loving public by storm. It produced no profits that might have enhanced the financing of the impecunious artist's impending trip, and instead left him on the hook for the printer's and binder's bill, which had been paid with borrowed money. As to who actually wrote the poems, the name Lucretia (Lucretia Walker Morse, that is) seems as good a guess as any.

The *Journal of Commerce* offered a modest stipend for the weekly newsletters, but Morse called the offer a distraction, and rejected it. He did, however, write a few pieces from Italy and France for the *Observer* during his year abroad.

The trip lasted for nearly three years. It must have been a dream come true for the man who had failed to make it to Paris in 1815. The outbound ship, a steamer named *Napoleon*, ended its voyage at Southampton, England. Most of December 1829 was spent in London, where Morse found his old roommate Leslie, now a confirmed expatriate, making a living as a portraitist.

In January, he was in Paris, and thence, via Lyon and Dijon, arrived in Rome. He was accompanied on the trip from Paris to Rome by a former neighbor from New Haven, Mr. Jocelyn, and a Mrs. Town of New York. His friend Cooper was waiting for him in the Eternal City, and the artist also found there a community of about two dozen expatriate Americans.

Morse spent a full year in Rome, during which he made a side trip to Naples and one to Florence. On the trip back to Paris, he detoured for a week in Venice. At this point, most of the twenty-two commissions from the trustees were completed; and all, including the few unfinished ones, had been shipped home.

The artist also shipped art books and plaster casts of famous statuary back to the Academy. An American in Rome, George W. Lee, donated a life-size copy of the "Farnese Hercules." Another, David Cohen, donated money for several smaller casts, including copies of anatomical models from Michelangelo's workshop. These additions doubled the number and variety of plaster cast models in the Academy's drawing studio. The books comprised the beginning of the Academy's art library. Morse packed all of these things carefully, and often accompanied them to Leghorn, the seaport near Rome. (He was once detained there for a day on suspicion of exporting stolen art treasures.)

The National Academy, meanwhile, was doing quite well as Gotham's only art institution besides Trumbull's moribund establishment. Its annual show, in March 1831, sold five-hundred "season tickets," and six-hundred at-the-door admissions. Thomas Cole and Asher B. Durand were the main drawing cards. Both of these young artists were virtually self-taught, and their distinctive styles had already been fully developed before they ever set foot in the Academy studios.

Durand owned a successful engraving business nearby. During his apprenticeship in Newark, N.J., he met Trumbull while engraving some of the artist's iconic Revolutionary-era scenes for a book. The veteran painter was impressed with his drawing and compositional skills, and suggested that Durand try his hand at painting with oils. He did, and it quickly became an avocation.

Cole, who emigrated from England while still a little boy, did not have any regular employment. He lived in a cramped furnished room with a small window that let in little light, and did his very large canvases there. Both he and Durand were "founders" of the Hudson River School—the first distinctly American movement or genre in any art form. Inevitably, they influenced many of the Academy's students. Frederic Church, perhaps the most famous of the Hudson River painters, exhibited some of his early work at the Academy, but spent little time there.

Morse left Venice just after the New Year in 1832, and reached Paris two weeks later. He roomed briefly with Horatio Greenough, the sculptor who would later design the Bunker Hill Monument in Charlestown, which was Morse's home town. The great challenge of his tour was at hand: a six-by-nine-foot canvas of a roomful of masterpieces at the Louvre Museum. The museum's practice in those days was to hang pictures one above the other almost from floor to ceiling, so the artist had quite a few to squeeze onto his canvas, which depicted three sides of the gallery. This fitted his purpose quite well, for his intent was to bring home as large a sampling of masterworks, few if any of them previously reproduced in the New World, as possible.

To that end, he did a painstaking miniaturization of thirty modest-sized paintings hanging on the one wall that the viewer sees head-on. The Mona Lisa was the smallest of the lot, in a gallery devoted to Italian, French and Dutch Renaissance masters. Morse faithfully included the doorway and the corridor beyond.

All this may seem unoriginal. But the idea of miniaturizing a large number of masterworks, in their authentic museum locale, was really quite unusual, although not without precedent. Even more daring was the artist's inclusion of not one but two self-portraits: painting at an easel by a side wall; and then, in a sideways view, hovering over his seated daughter Susan. She is front and center in the painting, drawing on a large sketch pad. His friend James Fenimore Cooper and family are also present, off in a corner, as spectators. A young woman (his late wife Lucretia, probably—or was it a gallery attendant?) is incongruously seated at a small desk on the open floor, reading a book. All of these figures lend the painting a certain "tension," being as prominently displayed as the paintings themselves.

The museum supplied him with a long work table, and a ladder for straight-on views of the pictures that were hung just below the ceiling. Work on this magnum opus went on for six full days per week over a span of two months. The busy pace somehow did not distract Morse from suddenly being seized by an urge to invent an electric telegraph, of all things. The incongruity was not lost on the many people who would challenge his patent in later years. "Gallery of the Louvre," not quite complete, accompanied the artist on his voyage from Le Havre aboard the *Sully* in October 1832. The first and most persistent of those challengers, Dr. Charles Jackson, was a fellow passenger.

Once home, the artist rented a storefront for paying admissions to see the Louvre painting. The opening weekend was well-attended, but then attendance dropped off drastically. The picture then travelled in New England for a brief while, with Thomas Prescott, the agent who had taken the House of Representatives on the first stops of its unsuccessful tour. Paid admissions, at least as reported by Prescott, failed to cover expenses.

A neighbor of James Fenimore Cooper's family in Cooperstown, NY, George Clarke, offered to purchase the painting in 1834. He gave the artist a note for $1300, payable in installments over three years. Clarke's payments always arrived very late, but they were a godsend for the artist, in what turned out to be a fallow period for commissions of any kind.

Shortly after returning to New York—and concurrently with his sudden fixation on inventing the telegraph—Morse secured a position on the founding faculty of the new City University—later renamed New York University. It was the first academic institution in the United States to offer art classes, and the position was clearly carved out for Morse, at the behest of the influential trustees of the Academy, several of whom were on the board of the new university.

It was not a salaried position, however. Students paid instructors directly, and tuition varied, depending on the agreed upon course of instruction. (By contrast, the Academy charged only a nominal membership fee, and had no regular faculty.) The school rented Morse a tiny round garret jutting out from the roof of its new Gothic-style building on Washington Square East (near Waverly Place) in Greenwich Village. The garret was a few steps up from his studio, and he was charged $100 per three months' rent for the garret and studio combined.

The artist taught at NYU for more than a decade—until 1843, when he moved to Washington to build the experimental telegraph. His telegraph slowly took shape in the studio between 1833 and 1837, and was introduced to the world in the ground-floor reception area in the latter year, to a large audience of invited guests.

In 1839, Morse was granted a leave of absence for a trip to England and France, in order to secure telegraph patents in those countries. Virtually nothing was accomplished along that line, but in Paris, he encountered Louis Daguerre and his Daguerreotype experiments. The father of photography allowed Morse to write down the mix of ingredients used in his plates and draw a detailed description of the camera that took the first photo ever, a street scene outside the inventor's window, in 1838.

Once home, and after Daguerre published a notice of his invention in August 1840, Morse had an exact replica of the original camera made at a local workshop. The shop also managed to make playing-card sized Daguerreotype plates. In September 1840, the camera was aimed out the window on a stairwell landing in the NYU building, facing east toward the Unitarian Church on Broadway. It took what historians believe was the first American photograph. The developed picture was blemished by silvery spots. This was the first photograph ever taken in the United States, but unfortunately the artist did not save it.[20]

A science professor, John W. Draper, helped Morse build a glass photography studio on the roof at NYU in October. One of the first of many photo portraits done there was a full-length image of Morse's daughter Susan, with some of her schoolmates. (Susan was then living with brother Richard's family.) They experimented with plate and camera improvements, and ways to enhance the lighting and reduce exposure time. (Morse's original photo of the Unitarian church required a fifteen-minute exposure.) Some of the country's first professional photographers, including Matthew Brady, visited the studio.

In 1840, Morse built another glass - roofed enclosure on top of the three-story building that housed his brothers' religious weekly, the *New-York Ob-*

server. The brothers loaned $500 to build it—and paid another $100, a year later, to have its remains removed after a violent thunderstorm.

With three students in tow, Morse took his camera obscura and fifty plates to Niagara Falls in 1840. The majestic setting was popular with the Hudson River School artists who frequented the Academy, but the trip did not inspire any drawings or paintings—just a series of daguerrotypes, shot from various vantage points.[21]

On 9 August 1841 he took a group picture of eighteen classmates from the Yale Class of 1811, in back of the president's house. (Seventeen members of the class were already deceased, and nineteen others did not attend the reunion.) It was the first class photo ever.

Morse abruptly stopped painting a few years later, for reasons that will be explained shortly. Yet he continued to be interested in photography as a hobbyist. The ex-artist, Professor Draper, his former colleague at NYU, were judges at the Crystal Palace Exhibition's photography show in 1853, and every year he renewed his membership in the National Photographic Association. Locust Grove, his home on the Hudson shore, contained a fully equipped photo lab. This we know from letters exchanged with Rev. R. L. Hill, an experimenter in color photography, during the 1850s.[22]

WHAT MADE MORSE GIVE UP ART?

In 1837, the year he first showed his new telegraph to the public, Congress appropriated money for murals in the rotunda of the Capitol. The committee that was put in charge of the project determined that the four spacious alcoves would be assigned to four different artists, of the committee's choosing. The award for each was $10,000, and the artists would have five years to complete their work, which had to depict an event in American history.

Morse, the painter of Monroe and Lafayette, as well-known as any artist in America, fully expected to be one of the designees. He was very keen on the prospect of doing a true history painting—especially one that would occupy so prominent a place in the very citadel of democracy.

A problem arose when John Quincy Adams, a member of the select committee, expressed the view that no American was up to the demands of such an important commission. He suggested that outstanding European painters should be selected instead. This provoked some controversy, and more than a few newspaper editors and letter writers stood up for American artists. An especially acidic letter that appeared over the pseudonymous signature "Junius," in the New York papers even questioned Adams's patriotism. It was fairly common knowledge that the true author was Morse's friend, James Fenimore Cooper.

Adams retreated from his position, saying it was only a discussion point, not a firm proposal. When the committee finally announced the four winners, they were indeed all Americans. But Morse was not one of them.

He was totally devastated. A letter to a close relative put it thusly: "The blow I received from Congress when the decision was made concerning the pictures for the Rotunda has seriously and vitally affected my enthusiasm in my art." He added that only the telegraph project saved him from "despair," and he further noted, "My prime is past, the snows are on my temples."[23]

Morse did not conceal the effects of this rebuff from others. A subscription was hurriedly scraped together by the trustees of the Academy, to fund a compensatory commission. Each kicked in a minimum of $100, and then they looked beyond their ranks for additional support. Among the donors were two leading Hudson River School artists and National Academy regulars, Thomas Cole (for $15) and Asher Durand ($30). Morse reluctantly accepted the award, which amounted to $2,000. He then announced his subject: a scene of the Mayflower Compact, titled "The Germ of the Republic" (not "Gem," as stated in one biography). It was the subject of the painting he had shown Washington Allston back in 1811, in Boston—the painting that launched his art career.

But Morse never even started work on this commission, and he began refunding the advances he had received, a little at a time. The depths of his despair were revealed when David Inman, one of the four winners and a long-time friend, dropped out for health reasons. The House committee invited Morse to replace him, but he turned the offer down.

Morse kept his post as head of the National Academy until 1843, the year Congress funded the Washington-Baltimore telegraph. The Academy had just completed an expansion of its building at Broadway and Leonard St., and stocked the added space with new easels and other studio accoutrements. An inventory counted seventeen large plaster statues, twenty-two smaller ones, assorted plaster heads, hands, etc., and a substantial library. In 1842, a summer exhibit had been added to the annual March event. Paid admissions in 1842 for the two combined were $4,000.[24] The ex-artist left behind a thriving institution, by far the best art school in the country.

Morse was succeeded as director by Asher B. Durand. In 1849, the Academy conducted a fund drive for a new and larger building. Durand and Charles Ingham gave $500 each; Jonathan Sturges (a trustee) donated $1,000. A total of $4,000 was raised as a down payment, about half of it obtained from $100 "subscriptions." (Among the subscribers was Frederic Church, now the most famous of the Hudson River School artists; and Morse, whose long years of poverty were just now coming to an end). The new building went up at Broadway and Mercer St., near Bleecker St.

In March 1861, Morse, then seventy years old, was visited by his old friends Durand and Ingham. The Academy was in dire financial straits, and the trustees were on the verge of cutting back its programs. Would Morse accept the presidency again, and try to straighten things out? He agreed to a one-year term. When the year was up, he acceded to an extension, until August 1863, with the proviso that there would be no more extensions. He was then succeeded by Daniel Huntington.

In 1865, somebody offered to sell the Academy a portrait of Washington Allston, painted by the inventor's former roommate in London, Leslie. Morse donated $500 toward the purchase price. He followed this up by donating Allston's last painting, "Balthazar," to the Academy. It had been given to him by the painter's widow, "with the paint still fresh."[25]

In 1868, Morse received a letter about a portrait he had once painted of the famous Danish artist, Thorwaldsen. It had been done in Rome, in 1831, as an "any subject" commission for Philip Hone, the former New York City mayor and one of the Academy trustees who had financed his trip. The current owner, who had bought the picture for $400, was interested in donating it to the Academy. Morse offered instead to buy it himself, explaining that he was about to embark for Europe, and wanted to make a gift of it to the King of Denmark. The painting's owner then simply gave it to the artist, gratis. Morse, as it turned out, did not visit Denmark, but personally delivered the painting to the Danish embassy in Paris. Along with the painting was a letter requesting payment of a telegraph patent royalty from Denmark.[26]

While on that trip, the ex-artist purchased several paintings in France and Germany by living artists, and sent them to the Academy for its next annual exhibit. This was unusual, because the Academy showcased only American art with rare exceptions for an occasional foreign visitor. If they sold, the proceeds would go to the Academy. If not, Morse would keep them.

The Academy was in a yet new location at this point, on the corner of 23rd St. and 4th Ave., not far from Morse's house (a second home, where he now lived from November to April each year). It had a daily schedule of live model drawing sessions, and regular classes in painting, sculpture, perspective, and anatomy. Guest lectures on art and art history were open to the general public, and the Academy also had an Antique School. According to its mission statement, "The schools will be open day and evening free of charge, the students furnishing their own easels and materials."

Advancing age and infirmity (he sustained a compound leg fracture in 1869) did not keep Morse away from the Academy's semi-annual art shows, his last being the 45th Annual Winter Exhibit, and the 21st Annual Summer Exhibit, in 1871.

In November that year, just a few months before he died, he was appointed to the official city welcoming committee for the visiting Russian Archduke Alexis. The highlight of the occasion was a ball, at the Academy of Music. A portrait of Admiral Farragut, naval hero of the Civil War, was unveiled by President Huntington of the National Academy of Design, with Morse standing at his side. It was a gift for Czar Alexander II. The artist was William Page, winner of a student competition that had been held for the occasion.

The long twilight of Morse's life in the arts included a few other highlights. As a founding trustee of Vassar College (in Poughkeepsie, NY, near his Locust Grove estate), he pushed successfully for a program in art, but turned down a request to be an examiner or judge of student art projects. He also endowed an art lectureship at Rutgers Female College.[27] In May 1871, he was named a charter trustee and vice president of the new Metropolitan Museum of Art in New York.

His last painting, a near-life-sized portrait of his teen-aged daughter Susan, still hangs in a place of honor at the Met. Only a few feet from where she sat for that portrait was an odd-looking contraption (not shown in the painting) that would change the world forever.

In December 1871, a neighbor asked Morse to paint his portrait, only to be informed that Morse did not have any paint or paint brushes, and hadn't painted a picture in thirty years.[28]

In 1849, Morse received a letter from his old friend James Fenimore Cooper, whom he hadn't seen in years. The famous novelist was broke, and wanted Morse to buy back his "Rembrandt" painting, which Cooper had bought for $600 a decade earlier. Morse told Cooper that, contrary to rumor, he was not wealthy, and in fact had no money to spare. His reply was very chilly, and surely he recalled Cooper's role in the Capitol murals disaster when he ended with this observation: "I am not a painter...I leave it to others more worthy, to fill the niches of art."[29]

NOTES

1. Samuel F. B. Morse Papers, Library of Congress, Manuscript Division, Diaries and Notebooks, 28 January, 1805.
2. Ibid., 30 January, 1805.
3. M to brothers Richard and Sidney, Morse Papers, LC, MS. Div., Letter Books, 12 August, 1810.
4. Cf. M—Parents, 15 May, 1812 (Letter File).
5. Quoted in William Kloss, *Samuel F. B. Morse* (New York: Abrams 1988), p. 26.
6. Ibid., p. 28.
7. Oliver W. Larkin, *Samuel F. B. Morse and American Democratic Art* (Boston: Little, Brown & Co., 1954), p. 31, sees the influence of the famous "Laocoon" in Morse's pose.
8. Ibid., p. 31. The painting was six-and-one-half by eight feet in size.
9. Kenneth Silverman, *Electric Man* (New York: Knopf, 2003), p. 15; Cf. also letter from Morse to Horatio Stone, Washington Art Association, 5 January, 1860; and Kloss, op. cit., p. 26.
10. M—Parents, 1 August, 1815.
11. M—Lucretia, 26 January, 1820.
12. M—Parents, 21 March, 1820; Silverman, op. cit., p. 51.
13. Ball—M, 16 March, 1821; M—Ball, 28 March, 1821.

14. Cf. Larkin, op. cit., p. 57.
15. Doolittle—M, 1 December, 1823; M—Lucretia, 19 February, 1823.
16. Sidney Morse (brother)—M, 29 October, 1824.
17. Later, "the Arts of" was omitted. Cf. Larkin, p. 85.
18. 1826 Letter File.
19. Cf. John Neagle—M, 15 May, 1828, on "attacks against you."
20. M—Fuller Walker, 14 August, 1871. The oldest surviving photo was shot by somebody in Philadelphia, a month later (October).
21. Robert Dodge (NYU student) "memo" dated 9 July, 1851. (1851 Letter File)
22. M—Hill, 25 January, 1855, for instance.
23. M—Edward Salisbury (cousin), 24 February, 1841. The "past his prime" artist was 46 years old.
24. M—Governor William Seward, 20 June, 1842.
25. M—Daniel Harrington, 4 April, 1865.
26. M—John T. Johnston, 23 January 1868; M—General Rusloff, 19 March, 1868.
27. M—Franklin Pierce, April, 1871 (no day).
28. M—T. Napoleon Cheney, 20 December, 1871.
29. Cooper—M, 17 November, 1849; M—Cooper, 25 November, 1849.

Chapter Two

Starving Artist Invents Telegraph in Greenwich Village Garret

There have been claims that Samuel F. B. Morse did not invent the telegraph at all, but stole other people's ideas. All of these accusations arose from the implausibility of a portrait painter mastering the intricacies of electrical engineering, in an era when there were no electrical engineers, to revolutionize the world of telecommunications.

The challenges were raised by venture capitalists and copycat inventors. Investors who wanted relief from the inventor's royalty demands often became the sponsors of the copycats. Others simply wanted some assurance that the rival claims that kept cropping up were spurious, before plunking down their dollars. Morse fought the challenges tooth and nail through the courts, one after another. And yet, the ultimate mystery remained: How did this down-at-heel painter transform himself into the great engineering genius of the age?

Morse's biographers have surprisingly little—practically nothing, actually—to say about this central issue, which provided the basic inspiration for the impostors and rivals whose lies preoccupied Morse until his dying day. The jarring juxtaposition of artist and inventor is a challenge for biographers, because it is not conducive to a smooth narrative (as the contrast between the present chapter and the one preceding will surely attest and illustrate). Imagine, if you will, a contraption made of wooden sticks and wire, taking shape in Morse's art studio on the top floor of the NYU building on Washington Square. Seeing it for the first time, a student, after concluding that it could not be some kind of anatomical model, would ask what it was. If he came back a year later, it would still be there in all its mystery, but looking somewhat different—probably more complicated. At the time the basic machine was completed, in late 1836, Morse was painting what turned out to be his last picture. It was a full-length portrait of his daughter, Susan, who stopped by the studio most weekdays after school, often bringing some of her schoolmates along. Looking at the large canvas, hanging today in the Metropolitan Museum of Art, there is no hint that the contraption was practically within arm's length of where she was sitting.

AN END AND A BEGINNING....THE END OF MORSE THE FAILED ARTIST; THE BEGINNING OF MORSE THE WORLD-RENOWNED INVENTOR

The aim of this chapter is to describe the making of the first telegraph, from its beginnings in 1832, to the initial public demonstration in 1837, and then the building of the first telegraph line, from Washington to Baltimore, in 1844.

According to the inventor's own account, the idea of the telegraph, and the broad outlines of how it should be constructed, came to him out of nowhere, during an ocean voyage. Morse was a well-known artist by then, and he was returning home to New York after spending a year in Europe painting pictures, mainly in Italy and France. The trip had been paid for by several of the National Academy's more generous patrons. All were Gotham businessmen, and Morse owed each subscriber in return a painting that would be executed during his trip. Some had specific requests, such as historic sites or copies of Renaissance masterpieces, while others left the choice of subject to the artist's discretion.

The sloop *Sully* was transporting about a dozen passengers from Le Havre. One day, the passengers began talking about Benjamin Franklin's famous kite experiments during electrical storms. The story of how Franklin drew lightning down a wire to pre-selected groundings (including, once, his own parlor, via an open window), was well-known. Morse and a fellow passenger, Dr. Charles Jackson, knew more about it than most. Morse had heard it explained scientifically in Professor Benjamin Silliman's classes at Yale, when he was an undergraduate there. Silliman was probably more familiar with the phenomenon of electricity that anybody else in the United States. He was the first American to write about the electric battery experiments of Volta and Ampere in Europe, c. 1800, and built a battery (probably the first in the United States) for use in classroom experiments. Morse recalled joining hands with fellow students in Silliman's class, and feeling a not unpleasant tingling sensation when the first student in this human chain grasped a wire sticking up out of the battery.

Dr. Jackson was a physician and also a geologist, who was professionally involved with electromagnetism, mainly in connection with the experimental use of magnets in prospecting for metals. He had just attended a conference on the subject in Paris, which was the home of several researchers in electromagnetism. Magnets, coils of wire and an electric battery were packed in his luggage.

Jackson's medical practice in Boston had somehow led him into geology. Some doctors, in those days, developed an expertise in botany, via their interest in medicinal plants. Others, including Jackson, used their knowledge of chemistry to study mineral-based cures. Minerals meant geology, a science that was just then coming into its own as a specialized discipline. He currently had a commission from the state of Maine to study rock strata for the purpose of locating mineral deposits, and would later hold a similar position in Michigan.

In Paris, Dr. Jackson attended seminars at the French Academy, where Professor Alphonse Pouillat and others described experiments with large magnets surrounded by wire coils that were energized by powerful batteries. Experiments in which acid and lead plate batteries (developed separately by Volta and Ampere) magnetized iron rods had first been carried out by Dominique Arago in 1820. Pouillat was a colleague of Arago who had assisted in his experiments. In addition to lectures and demonstrations based on Arago's work, Jackson was also apprised of similar experiments by Humphrey Davy in England, and Dr. Ernst Seebeck in Berlin.

Morse recalled asking Jackson if electrical sparks could be transmitted over distances. Jackson thought that they could. This reply, according to the recollection of several passengers (in sworn legal depositions years later) animated Morse with the idea that "intelligence" might be transmitted by means of a controlled spark. So animated was he (they recalled) that he would talk of hardly anything else for the remainder of the voyage.

Jackson acquainted Morse with the composition of the most advanced batteries, and when the artist suggested that sparks could burn coded patterns into paper, the doctor suggested that a spark might indeed make a legible imprint on chemically treated paper. He named some inexpensive compounds that might do the job. At journey's end, they exchanged addresses, but would never meet again. The year was 1832.

The artist began shaping telegraph transmitters and receivers out of wood and wire at his brother Sidney's house, almost as soon as he arrived in New York. He continued these activities after moving into a tiny garret at the newly-constructed City University (better known later as New York University). Thanks in part to the influence of his European trip sponsors, one of whom was Philip Hone, a recent mayor of New York, Morse had won an appointment to the charter faculty. His position as professor of Art was a first in American higher education.

The little corner turret in which he took up residence overlooked Washington Square. A short staircase was all that separated it from the spacious studio where he taught his classes and where he set up his telegraph lab.

After three years of constant effort, Morse had a functioning machine that he described in these words:

> My instrument was made up of an old picture canvas frame fastened to a table; the wheels of an old wooden clock, moved by a weight to carry the paper forward; three wooden drums, upon one of which the paper was wound and passed over the other two; a wooden pendulum first suspended to the top piece of the picture or stretching frame, and vibrating across the paper as it passes over the centre wooden drum; a pencil at the lower end of the pendulum, in contact with the paper; an electro-magnet fastened to a shelf across the picture or stretching frame, opposite an armature made fast to the pendulum; a type rule and type for breaking the circuit, resting on an endless band, composed of carpet binding, which passed over two wooden rollers, moved by a wooden crank, and carried forward by points projecting from the bottom of the rule downward into the carpet-binding; a lever, with a small weight on the upper side, and a tooth projecting downward at one end, operated on by the type, and a metallic fork also projecting downwards over two mercury cups, and a short circuit of wire, embracing the helices of the electro-magnet connected with the positive and negative poles of the battery and terminating in the mercury cups.[1]

He demonstrated it to students and colleagues in January 1836. The machine was hooked up to only forty feet of wire—and yet the signal seemed rather weak at the receiving end. Two members of the science faculty, John W. Draper and Leonard Gale, had looked in on Morse's work from time to time. Both knew something about electro-magnetism, and recent improvements in electric batteries. Gale was especially impressed by the possibilities of Morse's work, and began showing up nearly every day. Draper assumed the less exacting role of occasional scientific adviser.

Both Draper and Gale were acquainted with Joseph Henry, a physics professor at Princeton who had been endeavoring to build on the work of Arago, Pouillat, Davy, and Seebeck. Henry was arguably the world's leading authority on electromagnetism when, in 1831, he engineered an electric doorbell. This was the first practical application of the new science. Henry's bell, activated by a battery (lead plates immersed in acid) and an eighty-pound magnet, was not practical enough to be commercially viable. But Henry was more interested in science than in commerce. He measured the distances that electrical impulses of varying strengths would travel along a wire from battery to bell, and logged his findings. Draper had detailed notes from Henry's log, and he gave copies of them to Morse.

The battery setup in the studio became formidable, with multiple cups of acid. Batteries were still in an early stage of development, but the greatly improved configuration developed by the British scientist, J. F. Daniell, in 1836 was replicated by Professor Gale in the studio at NYU.

The magnet weighed more than one hundred pounds. It was surrounded by a dense layer of coiled and insulated wire—the insulation to assure that wire and magnet never touched. This cumbersome arrangement raised serious questions about the future of the telegraph project. What kind of gigantic apparatus would it take, for instance, to send an electrical impulse—a telegraph message—from one city to another?

Faced with a dilemma, Morse somehow stumbled upon a tentative solution. He described it thusly:

> to remove that probable obstacle to my success I conceived the idea of combining two or more circuits together...each with an independent battery, making use of the magnetism of the current on the first to close and break the second; the second, the third, and so on...A practical mode of communicating the impulse of one circuit to another...was matured as early as the spring of 1837, and exhibited then to Professor Gale, my confidential friend.[2]

Gale put it this way:

> One (1) battery at one terminus of a line of conductors representing twenty miles in length, from one pole of which the conductor to the helix of an electro-magnet at the other terminus (the helix forming part of the conductor); thence it returns to the battery end, terminating in a battery cup, O. From the contiguous mercury cup, P, a wire proceeds to the other pole of the battery; when the fork of the lever, C, unites the two cups of mercury, the circuit is complete, and the magnet, B, is charged, and attracts the armature of the lever, D, which connects the circuit of battery 2 in the same manner, which again operates in turn the lever, E, twenty miles farther, and so on.[3]

This was the electric relay, or combined circuit, which to this day is an essential piece of electronic hardware. Its invention, however, has been credited to Henry, in 1835. This would have been just in time for the Morse telegraph. Henry made a solid improvement in the design of electro-magnets, but there is some question as to whether the Princeton scientist actually originated the combined circuit. (This is discussed in detail in chapter 4.)

The "relays" that Morse put together were electromagnets positioned at intervals along a transmission line. An electric current, about to give out, would instead be jolted back to life at each relay, and continue its journey. It is safe to say that Henry's improved magnets were essential to the efficacy of the relay.

By this time, Morse was also well along in his efforts to develop an efficient system for marking (recording) electronic messages on paper. It is noteworthy that only a "recording" telegraph would do for Morse. The possibilities of a non-recording device with a receiving clerk taking notes from a moving dial, were never entertained for an instant. The inventor was apparently unaware that he had rivals in England and Germany. But none of them were taking on the added challenge of creating a recording telegraph.

Remarkably, the inventor had the idea for the Morse code from the very outset. He described it to his fellow-passengers aboard the *Sully* in 1832. Initially it involved patterns of dots with measured spaces in between that would equate to letters of the alphabet. In 1838, the spaces were replaced by dash marks. These allowed an infinite variety of dot-dash combinations, which made it easier to cover the entire alphabet and punctuation marks (also expressed in dot-dash) in a way that was clearly legible.

Recent literature has pointed out that the dot-dash system is the same as the zero-one system that is the basis of today's binary or digital communications. The idea of binary combinations can actually be dated back to the ancient Greeks, but the first serious elaboration of a binary system dates from Morse's lifetime—1854—when a mathematician named George Boole wrote about the infinite numerical combinations that could be derived from just two digits. Both work the same way, and enable an infinite number of combinations. In 1948, a mathematician named Claude Shannon, working at Bell Laboratories, determined the same thing on his own—he called the 0-1 pairing a "bit," and concluded that it was the basis of all communication. The subsequent programming of digital electronic appliances of all kinds, moreover, is fully compatible with the physical (positive-negative) properties of electro-magnetism.[4]

Morse had a volunteer assistant by this time—a student named Alfred Vail who was taking art classes to improve his skill at mechanical drawing. (Vail and Morse also attended the same Presbyterian Church.) It may have been Vail's idea to rank the letters of the alphabet from most to least used, by the simple expedient of visiting a few print shops, and counting the contents of their letter trays. (There was a separate tray for each letter.) The aim here was to assign the shortest dot-dash codes to the most frequently used letters. This would speed up both transmitting and decoding.

Vail lived just outside Morristown, NJ, next door to the family business, Speedwell Iron Works. Speedwell manufactured ship screws, railway car axles, and other heavy transportation equipment. The first telegraph transmitter and receiver were assembled there by Vail, assisted by his brother George. The brothers took it upon themselves to make modifications, as when Vail advised his teacher/patron that "I have dispensed with the large spiral wheel," and replaced it with a smaller one that did not crowd the magnet.[5]

On 3 October 1837, a "caveat" was filed with the U.S. Patent Office in Washington. This was a notice of intent to file for a regular patent in the immediate future, once some anticipated improvements were completed. Vail was sent to make the filing in person—at his own expense. He paid a $20 filing fee, and $45 for transportation, room and board.

Vail had also paid for the materials used on the Speedwell prototype. This curious arrangement was unavoidable, since Morse himself was virtually penniless. It was based on a contract, dated 23 September 1837. In it, Vail "covenants to construct and to put into successful operation at his own proper costs and expense one of the Telegraphs of the plan and invention of the party of the first part." Paragraph two of the document obligated Vail for the patent cost. Future improvements by either Morse or Vail or the two of them together were to be recognized "in proportion of their respective rights." This referred to a sharing of the patent, which the agreement identified as three-fourths for Morse, one-fourth for Vail. It was further stipulated that any foreign patents that Vail might secure on his own initiative would entitle him to a one-half share of any proceeds deriving therefrom.

In January 1838, a large number of invited guests gathered in the ground-floor parlor of the NYU building on Washington Square for the official unveiling of the telegraph. Several miles of wire were deployed around the room and into the foyer and nearby hallways, doubling back, and ending on a receiving machine set on a table close to the transmitter. The guests were served refreshments and given a brief guided tour of the arrangements. The demonstration, presided over by Morse and Vail, with Gale and Draper standing by, went off without a hitch. There was a robust round of applause, and several toasts to the inventor. Thus began the age of telecommunications. (No memorial to it can be found on the premises today.) A patent for the Morse Telegraph was issued in 1840.

New York was already the financial capital of the United States, and the inventor was obviously well-connected. But venture capital for the newfangled invention was nowhere in evidence. Morse himself was responsible for this. He regarded his telegraph as an express postal service, and hoped to sell his patent to the federal government for cash. The fact that Washington was then upgrading the postal system, with regular post offices and postmasters (and the first postage stamps) was encouraging. The telegraph would, in the inventor's view, be a vital part of that general upgrade.

A letter was sent to Treasury Secretary Levi Woodbury, offering to sell the telegraph, shortly before the caveat was filed at the Patent Office in 1837. In February 1838 (a few weeks after the NYU demonstration) Woodbury invited the inventor to bring his machinery to Washington. En route, Morse stopped at the Franklin Institute in Philadelphia and gave a successful demonstration to a large and enthusiastic audience. On 21 February, he met the Secretary at the Treasury Department and to his considerable surprise, found President Martin Van Buren and several members of the Cabinet in attendance for a demonstration. All were mightily impressed. Morse was told, however, that the sale of his device to the government was up to Congress.

Members of the House delegation from New York agreed to draw up a bill that included the inventor's asking price: $100,000. The bill was filed with the House Commerce Committee. Its chairman was Francis Ormond Jonathan "Fog" Smith, a businessman from Portland, Maine. Smith was quite accommodating, and sped the bill through his committee. The Committee's official report to the House noted that Treasury Secretary Woodbury was anticipating the adoption of a Federal telegraph, using Morse's invention. It continued that the telegraph "...will, in the event of success, of itself amount to a revolution unsurpassed in moral grandeur by any discovery that has been made in the arts and sciences, from the most distant period from which authentic history extends, to the present day."[6] (These words were written by Chairman Smith himself.) There was a cautionary note, however, that "it would be presumptuous...at this stage of the invention" to guarantee its success. Nonetheless, the telegraph was adjudged to be "...worthy to engross the attention and means of the Federal Government, to the full extent that may be necessary to put the invention to the most decisive test that can be desirable." The report concluded that, "...if successful...the Government alone should possess the right to control and regulate it." A $30,000 appropriation was requested (by unanimous vote of the Committee) for an experimental fifty-mile line, prior to any decision to buy the patent outright.

The Franklin Institute's official report on Morse's demonstration was similarly cautious. The Institute was quite impressed, but noted that the telegraph had not yet been put to the test of wires stretched across open country and subject to all kinds of weather. The relays—combined circuits, each connected to a battery and a bulky magnet enclosed in a watertight box—seemed especially vulnerable to mishap or mischief.

In a letter to Smith, Morse acknowledged that these large questions had not been answered. But he added that his invention had "immense power, to be wielded for good or evil," and also that it would make "one neighborhood of the entire country." Under these circumstances, he thought that only the Federal Government could develop the telegraph to its full potential.[7]

The committee chairman wrote up an authorization request for the $30,000 experimental telegraph. He tried to tuck it into an unrelated spending bill that was ready for a vote by the full House, but this maneuver was turned aside by the House leadership. He then announced that a personal emergency required his immediate attention back in Portland. The telegraph legislation languished on the House calendar, and expired when the 2nd Session of the 25th Congress adjourned three months later. This inaction should not have come as a surprise, considering that Congress had never appropriated funds for technological experiments before.

When the fall term began, the telegraph bill was again reported out of Smith's committee, and placed on the calendar of the House. Characterizing it as a sure thing, Smith decided that the new invention should be patented in Europe forthwith, before copycats began exploiting it there without paying royalties. Patent rights were something of a novelty in any case. The requirement of originality as a condition for obtaining a U.S. patent, for instance, dated only from 1836 (via passage of the Patent Office Act). Before then patents were issued indiscriminately.

He opted for another leave of absence from his House duties—this time to accompany the inventor to England and France in order to secure patents in those two countries. Just before embarking on their hurriedly scheduled trip, they received the additional news that a patent for a "Magnetic Needle Telegraph" had just been issued to Charles Wheatstone and William Fothergill Cooke in London. Under the circumstances, Morse was quite relieved to accept Smith's offer to pay all expenses for the trip.

Equipped with a complete telegraph apparatus, they headed straight for London, and appeared before the Attorney General, Sir John Campbell, with a patent application.

A lengthy interview was followed by a long letter from Morse to Campbell, outlining his claim to precedence via the presentation at the prestigious Franklin Institute in 1838.[8] When he returned for the official decision on his British patent request, Morse received a rude shock: his patent application had been disqualified. The reason given had no reference to the Wheatstone needle telegraph. Rather, it was the surprising news that the "details" of Morse's telegraph had been published in the *Mechanic's Magazine* (a London publication), the previous February, and thus his invention was now in the public domain.

It was learned, years later that the anonymous article had been submitted by Sir Humphrey Davy, who was also working on a telegraph. Davy had noticed a line about Morse's NYU demonstration in the "foreign bulletins" section of a London newspaper at the end of January. The squib he wrote for the magazine was a mere paragraph in length—a slightly wordier version of the bulletin and, like the bulletin, devoid of any technical details. Needless to say, it was the right honorable Sir Humphrey Davy who delivered a copy of the article to Her Majesty's Attorney General, the right honorable Mr. Campbell.

Campbell's decision merited a challenge in a court of law, but Smith was averse to the risk and the expense. The idea of a lengthy sojourn in London was a daunting prospect all by itself. The two Americans consulted a barrister, nonetheless, and he advised them that the British Parliament had the power to grant patents directly. The process, however, would entail significant legal expense, months of waiting, and no assurance as to the outcome.

Morse visited Wheatstone's erstwhile colleague, Mr. Cooke, who was kind enough to stage a demonstration of the "five needle" telegraph. Emboldened, perhaps, by the unwieldiness of that device, Morse decided to ask Justice Campbell for a rehearing. The request had to be made in person, and the inventor gave as the main reason for a rehearing the fact that the item in the *Mechanic's Magazine* was nothing more than a brief news bulletin, lacking in detail.

For good measure, he explained to the Attorney General the differences between his own machine and the Wheatstone-Cooke telegraph. The differences were clear enough: the British machine did not record, for one thing. Instead, arrows were activated by electro-magnets to point at letters on a dial. An attendant had to write down each letter as it was fleetingly pointed out. And of course there was no Morse code and, presumably, no means of transmitting messages over long distances. The rehearing was denied, with the explanation that no new facts had been put forward to justify it.

Before leaving England, Smith came up with the idea of approaching some wealthy Britishers with an offer to license the invention to them. Requests for interviews were sent to a select few, including Baron Rothschild, but prompt replies were not forthcoming.

Empty-handed, the two men set off for Paris. Smith all the while was keeping a meticulous record of his expenses, which included such diverse items as tickets to the coronation of Queen Victoria at Westminster Abbey, and a bill for mending a pair of "torn pantaloons."

In July 1838, the telegraph was demonstrated to a gathering of scientists at the French Academy in Paris. Among those present was M. Francois Arago, the pioneer of electromagnetic research. Arago was so impressed with what he saw that he arranged to address the Chamber of Deputies about it.

A French patent was applied for in October 1838, and duly granted. Early in the new year (1839) Morse was awarded the Great Silver Medal of the Academie de l'Industrie, along with an embossed diploma. Another telegraph demonstration took place at the Societé Polytechnique in January 1839. Shortly thereafter, a contract to build a private telegraph line along the right-of-way of the St. Germain Railway Co. was signed. It was to extend seven miles, between Paris and St. Germain.

But then came another rude shock. The contract was declared unlawful by a local magistrate, on the grounds that it violated the government's postal monopoly. The monopoly covered long-distance communications of every kind, and dated from the Napoleonic wars, when Paris was connected to all its boundaries via an elaborate visual "telegraphe." (The word was invented at that time.) The Chappe telegraph consisted of a network of hilltop stations equipped with towers from which moveable long arms extended. These resembled streamlined-looking windmills, or more exactly, modern wind energy towers (with the propellers not constantly rotating). The signals followed the same principles as ship-to-ship communications, which depended on flags held in various arms-length combinations by the signalmen. In clear weather, simple messages could travel nearly fifty miles an hour. News from the Mediterranean coast could reach Paris within three hours. There were obvious national defense implications in all this, and the central government wanted to retain direct control over it. It was the world's most extensive telegraph system.

Morse's immediate response to the private telegraph veto was to offer his invention to the French government, for roughly the equivalent of the $100,000 he was asking from the U.S. Congress. Arago was willing to approach members of the Chamber of Deputies on the matter, but any purchase bill would have to wait for the next session of the legislature.

Smith was no more prepared to pay for a sojourn in Paris, awaiting an uncertain outcome, than he had been to await the deliberations of the British Parliament in Westminster. He wanted to go home.

Hopes for an English patent were revived, meanwhile, when Lord Elgin and the Earl of Lincoln approached the two Americans following the demonstration at the Polytechnique. They were very impressed with the telegraph, and promised to pull strings with their colleagues in the House of Lords, in order to move a patent bill through Parliament.

The inventor was also approached by the Russian ambassador in Paris, Baron Peter Meyerhoff. The Russian even went to the trouble of having a duplicate of the demonstration model made at his own expense, after securing the inventor's permission. He shipped it to St. Petersburg for an inspection by Czar Nicholas I.

At this point, Smith was counting his pennies. The two men accepted an invitation to stop in London, en route home, and put on a demonstration for potential investors at the Earl of Lincoln's house in Park Row. This event took place on 19 March 1839. Two weeks later, they were riding the waves of the Atlantic, still with high hopes, but with no clear-cut achievement in hand.

Nothing came of the British patent project, as the efforts of the two British lords to obtain a patent were blocked. Wheatstone secured a British patent for an improved version of his device in 1840—just a month after Morse's U.S. "caveat" finally eventuated in a U.S patent. Wheatstone telegraphs began operating between Liverpool and Manchester in 1841, before there were any Morse lines in the United States, and a U.S. patent for the British telegraph was applied for, and granted in 1841. Also in 1841, a German named Franz Steinheil secured a patent in his native Bavaria for a distinctly different telegraph machine, and built a telegraph line from Munich to an outlying suburb, along a railroad right-of-way.

Nothing was heard from Meyerhof for many months. Finally, in response to an inquiry sent in August 1839, the Russian diplomat explained that a projected line between St. Petersburg and Warsaw had been vetoed by the Czar, who was concerned that the ever-rebellious Poles would vandalize it. Meyerhof did not explain why no other routes were considered, but he did say that the Czar was very impressed with the machine.

All of these frustrations were nothing in comparison with what awaited Morse and Smith upon their return to the United States. The second telegraph bill had made it from the House to the Senate calendar, but the session ended before it came up for a Senate vote. Smith's "sure thing" had not been so sure. To make matters worse, Smith had been pressured into resigning his seat in the House, when it became clear that his stay in Europe would stretch out over several months. Friends in Portland advised him, upon his return, that it would be bad form to challenge his replacement in the next (1840) election. Due to the rules of seniority, the replacement had no influence in the House at all.

Returning to his teaching job at NYU, and still broke, the country's only college professor of art was faced with a bill for seven months past-due rent on his large third-floor studio and tiny bedroom in the corner garret. Such was the financing in Academe in those days. Students paid tuition not to the school, but directly to their professors, who then paid for the facilities. He promised half of his future tuition receipts toward catching up on the rent, and the school allowed him to stay for the time being.

Eager to resuscitate his fortunes in Congress, the inventor wrote letters to members of New York's delegation in the House and Senate, but received in reply only friendly notes of encouragement. He heard nothing from Smith, who was engrossed in his newspaper and real estate business in Portland, and Vail was back in New Jersey, working in his father's foundry. Professor Gale's interest in the telegraph during this period of stasis was, by contrast, unflagging. His chemistry lab was just downstairs from the studio, and most days he would come bounding up the steps for a brief session of tinkering with the batteries and relays.

In June 1841, a Washington lobbyist named Isaac N. Coffin contacted the inventor. He said the main obstacle to an appropriation was the element of risk—betting on the unknown, as it were—and the considerable cost. A semaphore system was currently being tested by a Louisianan named Gonon from the top of the Capitol building. Based on visual signals rather than complicated wires, magnets and batteries, it was being demonstrated at no cost to the taxpayers, and had the support of Louisiana lawmakers in both houses.

Coffin expressed confidence in the superiority of Morse's invention, and thought he could win congressional support for an appropriation. A commission of 10 percent was asked for, but Morse stood firm on 5 percent, and Mr. Coffin reluctantly accepted. This lobbying effort produced no prompt results, and in December 1842, the deal was terminated.

During this interval, Morse and Gale experimented with wire insulation. Wire resting on river, lake and bay bottoms presented the greatest challenge. These efforts proved especially frustrating. There seemed to be no 100 percent reliable insulating material, and no reliable mechanical means for applying it. In addition, there were physical hazards—as when a passing boat dragging a net cut a line that the two men had managed to stretch all the way from Castle Garden to Governors Island, at the southern tip of Manhattan. (This happened in October 1842.)

Samuel Colt heard about these endeavors, and did a similar experiment across the East River. Colt was the inventor of the repeating firearm, but like Morse's telegraph, it hadn't yet captured the public's imagination or attracted any venture capital. Colt's interest in telegraphy was limited to this one episode, and Morse was there to assist him. The wire had a *gutta percha* coating, which was the best thing available—and which, in fact, would be used on the trans-Atlantic cable years later.

The inventor's British rival, Wheatstone, obtained a U.S. patent for his non-printing telegraph in 1841, a year after his British patent was granted. He contracted with a Philadelphia machinist to make prototypes for franchisees, who would then be allowed to make copies as their business grew. Wheatstone's "needle telegraph" of 1837 had been renamed the "electro-magnetic telegraph"—a term that Morse had originally used for his own invention. In reply to a request from the British inventor, the Franklin Institute determined that Wheatstone's invention predated Morse's. This conclusion was based on a reprint in the Institute's Bulletin dated 19 May 1837, of an 1836 article on Wheatstone, which had appeared in the British journal, *The Magazine of Popular Science*.[9] Morse's filing with the Franklin Institute was dated 8 February 1838. The Wheatstone article actually described a work-in-progress, but the Institute considered it a work that was complete in all its essential aspects.

Morse was quite upset by all this, but when he examined a copy of Wheatstone's U.S. patent application, he was somewhat relieved to find that his rival's machine still did not record data, and continued to rely on an unwieldy array of moving dials, operated by a veritable maze of wiring. It was the same instrument Morse had seen in Cooke's workshop, during his sojourn in London. Yet Wheatstone's machines were already deployed in the English Midlands, along the right-of-way of the Great Western Railway. This was cause for concern.

Morse finally decided to take matters in hand and go to Washington to personally lobby Congress, as soon as the fall term ended at NYU. He managed to borrow a small sum of money from Gale and Vail to cover expenses, but such was his desperation that he told them he would not come home until the issue was decided one way or another—even if the money ran out, and even if NYU did not grant him a leave of absence from the spring term.

At the time of his arrival in December 1842, Gonon's semaphore was still under consideration. Right after the New Year, House members from New York State began arriving for the new session of Congress. Morse met with all of them. By the time the session got underway, a telegraph purchase bill was ready in draft form for submission to the Commerce Committee. Like its predecessors, the bill requested $30,000 for the construction of a trial telegraph between Washington and a nearby city. Alfred Vail came to Washington in January to help set up a demonstration that included connections between congressional offices and committee rooms in the Capitol.

The bill passed the House Commerce Committee—Smith's old bailiwick—in February 1843. It next went to the Ways and Means Committee, which was chaired by Millard Fillmore, the senior member of the New York State delegation. Fillmore pushed it through in a few days, and sent the measure to the full House with priority status, which jumped it ahead of a logjam of other bills awaiting action. The full House approved the telegraph bill, 89 to 83—less than a week later. The close vote followed a brief but pointed debate on the floor of the House that included remarks about wasting money on crackpot schemes. One lawmaker only half-seriously offered an amendment requiring that part of the $30,000 be used for a study of mesmerism. (The amendment was defeated.)

The Senate proved to be a tougher venue. The bill cleared committees, but received no precedence on the long backlog of spending measures that were awaiting action. The postmaster general, Cave Johnson, refused to support the measure. *The New-York Day Book*, a penny paper, noted that Johnson "abusively, derisively, bitterly" opposed the telegraph grant. This was a serious matter, since the whole idea of this legislative push was to make the telegraph a part of the postal service.[10]

As the session neared adjournment, the telegraph bill inched its way up the Senate calendar, still some distance from the top. Morse at this point would have been sleeping in doorways had he not found a room available at a boarding house owned by his old Yale classmate, Henry Ellsworth. Better yet, Ellsworth was very accommodating on the room and board fees for his old school pal. Remarkably enough, he was also the U.S. Patent Commissioner—the very man whose signature was on Morse's 1840 patent, and on Wheatstone's recently granted American patent as well.

On the day before adjournment, Morse packed his carpetbag and went down to the station to buy a ticket back to Gotham. NYU had granted his request for a leave, so at least he had a job of sorts to go back to. Seats were available only for the next day. He bought his ticket and, with less than a dollar in his pocket (thanks to another small loan from Vail), returned to the rooming house.

Late the following morning the Commissioner' teenage daughter, Annie Ellsworth, found Morse sitting on a bench in the rooming house foyer, his carpetbag on his knees. He was about to leave for the stagecoach depot. "I've just come from the Capitol," Annie said, "and I have good news."

As much as he despised the telegraph project, Postmaster General Cave Johnson did nothing to scuttle it, and proved easy to work with.[11] A contract was drawn up, setting out the bookkeeping and reporting requirements in detail, and providing for regular salaries to Morse, who was designated as superintendent, and his designated assistant superintendents, among whom Vail was second-in-command. Morse was allocated $2,000 for one year, paid by the week, and Vail $1,400. A newcomer named Henry J. Rogers was hired for $1,000 to set up and run an office in Baltimore, which had been selected as the terminus of the trial line. Rogers had come out of nowhere asking for the job. His background in railroad management (for a local line based in Richmond, VA) seemed apropos, since the lines were to be installed along the B&O right–of–way, all the way to Baltimore, where Rogers now lived. Pay for telegraph clerks was set at $300 for the year, with the expectation that six would be hired.

"Fog" Smith hurried down from Portland, bringing with him a young man named Ezra Cornell. His new acquaintance had read about the telegraph project, and Smith's association with Morse, in a newspaper. Setting out from his home in Ithaca, NY, Cornell hitched wagon rides and walked all the way to Portland. His aim was to sell an entrenching machine that he had recently invented, having read that the wire was to be buried in the ground. On the strength of Cornell's mechanical drawings, Smith paid $100 to have a model built. The horse-drawn machine dug an impressive trench in rocky, hard ground on Smith's farm at Forest Home, just outside Portland.

A major expense item for the project was the five tons of copper wire and seventy thousand yards of lead pipe casing for it that were ordered from a Manhattan foundry. The wire cost $3,400, the lead pipe $4,950. This was 20 percent of the total budget for the year, and did not include transportation. It far exceeded the estimate Morse put into his approved budget. That estimate was pure guesswork, before Professor Gale was enlisted to find a supplier and obtain an actual price.

Work got underway in autumn 1843, and went on through a very cold and blustery winter. Cornell operated the entrenching machine, which was shipped down from Forest Home. He then had another one built—and a new operator hired and trained—when both he and his machine began showing signs of wear and tear. When the ground froze in January 1844, the machines gave way to men with pickaxes and shovels. The additional machine and the resort to manual labor put further strains on the budget. But the work went on apace.

It was during that freezing month of January, however, that a major problem arose: short circuits in the freshly laid and buried wire. An NYU professor named John Fisher had been hired to inspect the completed lead pipes (which came in three-hundred-foot-long sections), and the coils of wire, prior to shipment from Manhattan. Morse became suspicious about the care with which Fisher was inspecting the materials. His concern was heightened after an inquiry was sent asking the manufacturer why deliveries were always late. The reply placed the blame entirely on Fisher, whose inspections were few and far between. On being asked about this, Fisher simply said that he had a busy schedule and was not available every day...or every week, for that matter.

Morse became convinced that Fisher was the source of the alarming regularity of lines suddenly going dead, which threatened the success of the project. Smith, who was being paid a modest salary to prepare regular reports to the postmaster general and keep the accounts in order, was quite upset by the technical problems, which he quickly decided were the inventor's fault. He confronted Morse on the issue of never having tested his invention in a real world environment—outside a studio or somebody's parlor. As the disruptions continued, Smith accused the inventor of using Fisher as a scapegoat. Next came threats: a lawsuit against Morse for a refund of the $3,000 he had spent on their European trip, plus amends for having been lured away from a promising political career, in the interests of pursuing this pig-in-a-poke. Smith became livid with rage when Morse coolly replied that he had no money to pay damages, and that if the project turned out to be a total fiasco, the government would surely zero in on Smith, who had been granted a one-quarter interest in the patent on the occasion of the European trip.

According to Ezra Cornell, in a recollection of the Washington-to-Baltimore for Morse's first biographer, Samuel I. Prime, this episode was the source of the litigious and bitter enmity that would persist between these two principle owners of the telegraph patent from that time until the inventor's death in 1872.[12]

Toward the end of January 1844, with the source of the problem still a mystery, everyone agreed that the wire would have to be strung above ground if the project was to have any chance of success. The Baltimore & Ohio Railroad, which provided the right-of-way, had no objection, so Morse immediately sent rush orders for twenty-foot poles to several lumber companies. The lead pipes and the wire inside them were all disposed of the following July as scrap, at ten cents on the dollar.

Labor costs for the two-dozen-or-so workmen who were always on the job averaged $450 to $500 a week. Cornell had signed on for a fixed fee of $800, which remained the same even though half his work consisted of digging up the rejected pipe. While doing this, he discovered the real problem with the outages. It was the heat from soldering the pipe-ends together, which damaged the wire where the heat was applied.

Morse gained some relief from his budget problems by refusing to remit any of Professor Fisher's $1,000 fee, after dispensing with his pipe and wire inspection services. Fisher didn't take this well, especially since he was being blamed for the short circuits. He sent threatening letters, but there the matter eventually ended.

Cornell was put in charge of installing the overhead lines. This endeavor was completed with impressive speed—and no line failures. The telegraph project had been rescued. Remarkably, considering all the unanticipated expenses and the fits and starts through the month of January, the job was completed not only ahead of schedule but also $3500 under budget.

A telegraph was set up in the unlikely confines of the Supreme Court Building, in Washington. Most of the justices were present, along with various senators and representatives, the former First Lady, Dolly Madison, and the patent commissioner, Henry Ellsworth. Annie Ellsworth, his teenage daughter, had been designated to provide the first telegraphed message ever. She settled on a passage from the Scriptures, which reads in full: "Surely there is no enchantment against Jacob, neither is there any divination against Israel; according to this time it shall be said of Jacob and of Israel, What hath God wrought." It was an ambiguous passage, the product of a stiff translation into seventeeth-century English. But there was nothing ambiguous about those last four words, which comprised Annie's message.[13]

NOTES

1. Morse deposition in Morse, et al., v. O'Rielly. Southern District Court, Federal District of Kentucky, Frankfort, KY. 1848 Letter File.
2. Ibid., Cf. also Morse note in 1836 Letter File (August); and M—Gale, 11 May 1869.
3. Gale deposition, Morse et al., v. O'Rielly.
4. Cf. Arpad Barna and Dan Piorat, *Integral Circuits in Digital Electronics*, New York: Wiley, 1973, for a clear and detailed exposition of this subject.
5. Vail—M, 14 October, 1836.
6. United States House of Representatives, House Commerce Committee, "Official Report on a Telegraph," Washington, D.C.: 1838. Copy in Morse Archive, Miscellaneous File.
7. M—Smith, 12 March, 1838.
8. M—Campbell, 12 July, 1838.
9. No. 5, p. 466.
10. Miscellany Folder, Morse Papers.
11. His allegedly negative viewpoint was reported in the New York Day Book, a penny press daily, on 8 December, 1843. A clipping of the story is in the Morse Letters folder for 1843.
12. Cornell–Prime, 24 May, 1874. Prime was gathering material work on a biography, commissioned by Morse's wife, Sarah, and inserted the Cornell letter into the Morse papers.
13. Numbers, 23.

Chapter Three

From Wilderness to Empire

Morse and the System Builders

The telegraph made front-page news again the day after its inauguration. On that date, delegates at the Democratic Party convention in Baltimore nominated James K. Polk for president, and Senator Silas Wright of New York for vice president. The news was telegraphed to Polk and Wright back in Washington within minutes. Wright, apparently taken by surprise, decided that he did not wish to give up his Senate seat. He handed the telegraph messenger a note to that effect, and his rejection of the nomination was telegraphed and read to the Convention less than an hour later. The delegates telegraphed a plea to reconsider, but Wright telegraphed back with a firm "no."

The next day's newspapers gave as much space to the "medium" as they gave to the "message" (to paraphrase Marshall McLuhan, in his famous book, *The Medium Is the Message*). They noted that the entire exchange was accomplished with lightning speed, and that the delegates then had ample time to select another nominee for vice president before the day was over.

After this spectacular beginning, the Washington-to-Baltimore telegraph settled into its day-to-day operations. Congress had appropriated, in addition to the construction funding, an operating budget of $8,000 for a single year. The aim was to see if the telegraph could stand on its own, financially, by the end of that period. The fact that there was just one telegraph office in each of the two cities was certainly a drawback, although the central location, in each city's main railroad terminal, was a plus. The fact that there was no network, but just a two-city connection, was a major handicap. Had they been connected to a multi-city network, the two offices would have had a much greater chance to pay their way and earn a profit.

Matters weren't helped, either, when Morse decided to double his asking price—from $100,000 to $200,000. He did say that this was open to negotiation, but since the appropriation for the experimental line had been passed by Congress on the premise of a $100,000 purchase price, this looked like double-dealing. The inventor had no reason to believe, yet, that the Washington-Baltimore line would be successful enough to justify even a $100,000 sale price. But he was in an optimistic mood, and having awarded nearly half of his patent right to Smith, Gale, and Vail (collectively), was now concerned about the size of his own reward.

The optimism didn't last long. Revenues fell far short of costs, and the government's operating budget was exhausted in just over six months. An additional $4,000 had to be appropriated in order to keep the line open. Business remained slow thereafter, but on a modest week-to-week upward trend.

Morse hired a Washington lobbyist named Amos Kendall to promote the sale of the patent to the government. The fee was a 10 percent commission on the sale price. Kendall was exceedingly well-connected with the Democrats, who controlled the White House and both houses of Congress. He had served as postmaster general during the recent Van Buren Administration, and so could speak with authority on how well the telegraph would fit into the postal system. Kendall also had a reputation for honesty, which was based on the fact that he appeared no wealthier at the end of his public service than he had before. Indeed, he was very frank in his first meeting with Morse, as to what he could and could not do: the telegraph would ultimately have to sell itself, he advised.

Smith proposed a scheme for making the telegraph more attractive to the government. It involved the formation of a private company that would make a commitment to lease the line, after it was bought by the government. The company would post a security deposit, and promise to extend the line to Philadelphia, the country's second largest city, at its own expense. The government would be given an option to buy this extension, which initially would belong to the company. The company, in turn, would consider buying up the government's stake at a generous mark-up, based on any revenue enhancements that resulted from the extension to Philadelphia.

Smith followed up by writing letters to prominent citizens in the three cities. Only a Baltimore businessman named David Burbank showed any interest. He was just then disposing of the leftover lead pipe from the aborted underground line, which he had bought for pennies on the dollar. (For this, he may be considered the first person ever to make any money off the telegraph.) Burbank promised to invest $10,000, provided he could draw some of his business acquaintances into the project as partners.

Kendall meanwhile got a bill started in Congress to buy the patent, even though the trial period for the experimental line was barely two-thirds over. It passed the House Commerce Committee, won approval by the full House, but died in the Senate when Congress adjourned. The trial year then expired, with the U.S. Government in possession of an unprofitable telegraph line. In May 1845, Morse, his patent partners, Mr. Burbank and a few Baltimore investors founded the Magnetic Telegraph Company (MTC), with headquarters in Baltimore.

The government agreed to surrender its property to the new company for a token payment. This largesse helped to attract several new investors. Amos Kendall, the lobbyist, took the credit for it, and also assumed the management position at the MTC's Washington office. He immediately began planning to build a line from Washington to New Orleans.

Morse was deeply troubled by the government's failure to buy the patent, for at least two reasons. One, not being a businessman, he just wanted a clean break from the world of business. His experiences of the past year, as superintendent, somehow did not awaken any entrepreneurial instincts. Morse only wished to bank his patent share and move on with his art career. The other issue was what he perceived as the need for a centrally controlled, integrated telegraph network. It was a natural for the postal system, and no alternative to that was anywhere on the horizon.

The lack of profits on the Washington-Baltimore line weakened the telegraph's appeal to venture capitalists. Matters were not helped by the fact that the Magnetic company was saddled with a patent royalty amounting to 25 percent of net profits (down from an initially agreed 50 percent). Ezra Cornell, the builder of the experimental line who was now the construction manager for the Magnetic, was the only persistent optimist: he actually borrowed money ($500) to buy company stock, and pledged a quarter of his salary for the purchase of more.

Enough money was scraped together, finally, to build the extension from Baltimore to Philadelphia. Once this was completed, new subscriptions began trickling in, and construction continued to the Jersey shore of the Hudson River, opposite Manhattan. There, Morse, Vail, Cornell, et al., pondered anew the challenge of traversing water with an electric wire. They had forded numerous small rivers and streams with overhead wires. But the Hudson was something else entirely. Its great width, strong tides and currents, and heavy traffic that included high-masted ships that overhead wire would not easily clear, presented a daunting prospect.

At that point, a twenty-nine-year-old man from Rochester, NY, named Henry O'Rielly contacted Morse with an offer to build a line from Philadelphia to Pittsburgh. (Note: "i" before "e" is the correct spelling, although he legally changed it to the more familiar "ei" in his old age. The O'R archive at the New-York Historical Society, formerly listed under "O'Rielly," is now listed under "O'Reilly.") O'R was an Irish Protestant immigrant who had arrived in Rochester with his parents at the tender age of nine. He was a printer by trade, and his newspaper, the *Rochester Democrat*, was the only daily in upstate New York. (Rochester itself was then a metropolis of thirty-five thousand people.) This made him an important voice of the Democratic Party in that region, and for that reason, he had served as postmaster of Rochester (appointed by Postmaster Kendall) until 1841, when the Democrats lost the White House.

It may have made a difference that "Fog" Smith was a Democrat, too; and Morse, for whatever reason, was an especially devoted Democrat. The young man out of nowhere offered to find his own investors, and have a line over the rugged Allegheny Mountains and into Pittsburgh in half a year. Morse and his patent partners accepted this offer, on the condition that once O'R made it as far as Harrisburg, he would have to pass an inspection of his construction and finances before being given clearance to Pittsburgh. In return, if he got this clearance, he would have first-refusal rights on all construction further west. O'R told his newspaper readers back home that, "The editor of the Rochester Democrat must shortly straddle a streak of lightning, so as not to be behind the intelligence of the age."[1]

Two months into this westward branch line, as O'R's workers passed Harrisburg, Vail and Gale went out on an inspection tour. They were disappointed by what they saw. Vail found that O'R was using glass insulators on the telegraph poles that were designed by Cornell, rather than his own, which were thicker and more expensive. Wires were slack and often too close to the ground, and poles were tilted.

Within days, the Irishman from Rochester was notified that his contract was terminated. He did not reply immediately, and his work crews continued stringing wire and erecting poles across Western Pennsylvania. Morse filed for an injunction in the Philadelphia Federal District Court. Rather than issue an immediate order, the judge scheduled a hearing. In court, O'R made a counterclaim that Morse et al., had negotiated his contract in bad faith, citing its vague wording as evidence. The document contained no specifications as to quality, and no firm deadlines for completion. All it said was, "good construction...must reach Harrisburg in a timely manner." The patent holders knew that he could not afford to remove and dispose of the poles and wire he had stretched halfway across Pennsylvania. Therefore, he asserted, they were confidently expecting that the installation would simply be abandoned, so they could then take it for free.

Morse countered that he'd prefer new poles and wire and a better right-of-way, and then raised the issue of an entity called the Atlantic, Lake & Mississippi (AL&M) Telegraph Co. The contract conferred ownership of the new lines on the Magnetic Telegraph Co., so what was this new company all about? O'R replied that he had created it as a fund-raising vehicle. Investors bought shares in the AL&M, and the money was used to build the branch line. The judge asked what investors in the AL&M then actually owned, if not the branch line itself. The man from Rochester replied that they were buying an exclusive lease from the Magnetic company. This was news to Morse, who thought somehow that O'R would be the sole lessee. He complained to the judge that O'R should have been more forthcoming about his ultimate plans. The man from Rochester replied that nobody could possibly expect him to go it alone, especially when his contract included a grant "beyond Pittsburgh to any point of commercial magnitude." This, he added, made a great impression on his investors.

The judge then wondered exactly who O'R had made a deal with. Was it the Magnetic company? None of the company's officers had signed the contract—which bore the signatures of the four patent holders and O'Rielly, and nobody else. The judge turned down the injunction request.

Morse and his partners then published ads in the local newspapers, announcing their repudiation of the O'Rielly contract. O'R then notified them that the deal was still in effect, and he threatened legal action to enforce it. Meanwhile, it was full-steam ahead in Western Pennsylvania. The branch line was in Pittsburgh a few months later. The grand openings of telegraph offices in the Philadelphia and Pittsburgh train depots, and in Harrisburg and the other towns along the way, were announced with full-page advertisements in the newspapers. Refreshments were served, speeches were made, and the Atlantic, Lake and Mississippi Telegraph Company was on its way.

Right after the court hearing, "Fog" Smith made an offer to his patent partners. If they would simply transfer the O'R contract to him, "beyond Pittsburgh to any point of commercial magnitude" and all, he would assume full responsibility for dealing with the upstart from Rochester. They readily agreed, and all signed on the dotted line when Smith pulled a contract out of his pocket.

In the ensuing months, Morse, Vail, and Gale became increasingly mystified by the fact that this new deal was having no discernible effect on O'Rielly. His AL&M was going full blast, branching out north, south and west from Pittsburgh. Smith filed a suit in Pittsburgh against O'Rielly but then asked for a continuance, and then another. Meanwhile, he didn't plant any poles or string any wire.

Remarkably enough, Smith had a brother-in-law who happened to be one of Pittsburgh's most prominent bankers. He never let on about this to Morse, who found out about it only years later, via a private investigator he hired to pry into Smith's affairs. The banker, Eliphalet Case, happened also to be one of O'Rielly's principal backers, and this relationship did not change when the partners transferred the O'R contract to Smith in 1845.

The Smith archive contains a dense correspondence between the two in-laws, indicating that they were on very cordial terms throughout this era. Smith was very close to his sister, Helen, who seems always to have been in delicate health. He visited the Case household every year or so, for a couple of weeks stay.

Later in 1845, O'R incorporated the Atlantic & Ohio Telegraph Co. It was a subsidiary of the Atlantic, Lake & Mississippi Company, and connected Pittsburgh with the Great Lakes. Case was named president of this company.[2]

Morse was distracted from this unfolding fiasco by a more favorable development. Some businessmen from the Utica-Ithaca-Albany area of New York State approached him about a patent right for a line that would link Gotham with Buffalo, via Albany. This presented him with a difficult choice, because the offer was for a company that would be totally independent of the Magnetic Telegraph Co. The inventor was still clinging to his dream of a monolithic network, which seemed the sensible business model for his invention. He was also still pondering how to get his wires across the Hudson from New Jersey to Gotham. Ezra Cornell, who was from Ithaca, was one of the drivers of this project. On securing a commitment for a 25 percent royalty, plus the assurance that the ever-reliable Cornell would be the construction manager, Morse gave in. His patent partners went along without demur.

In its first year of operation (1847–1848), the New York, Albany & Buffalo Telegraph Co. (NYA&B) provided Samuel F. B. Morse with the first substantial rewards from his great invention. By contrast, the Magnetic line continued to struggle with excessive operating costs, unreliable service, and low traffic. Kendall's New Orleans project crossed a vast rural landscape in order to reach the Crescent City, and so its profit margins would remain stubbornly modest. (They would disappear entirely for the duration of the Civil War.) The outlook west of the Alleghenies was "fogged" by the machinations of "Fog" Smith and his friendly enemy, Henry O'Rielly. Neither was involved in a new company that was founded in Louisville, in 1849, that had such solid backing from local businessmen and bankers that Smith decided the prospect of patent royalties trumped any inclination to throw obstacles in its way.

This pattern of telecom fiefdoms was not to the inventor's liking, and everybody knew that it would someday give way to a more standardized, centralized business model. But in the meantime, "Fog" Smith appeared destined to become one of the principal fief magnates, thanks to his assumption of the defaulted O'Rielly contract. A federal judge in Philadelphia, however, had cast a shadow over its legality by denying a cease and desist motion against O'Rielly. The judge doubted that the patent holders had the right to issue the contract, as the self-styled representatives of the Magnetic Telegraph Co. However, he held back from declaring the deal null and void, pending further examination—which none of the litigants showed any eagerness to pursue.

This uncertainty was the basis of Smith's arms' length relationship with O'Rielly, via Smith's brother-in-law, Eliphalet Case. Determined to carve out a telegraph fiefdom that he could call his very own, his hopes then settled on the part of the country that he knew best—New England. The major attractions there were twofold: one, a line connecting New York and Boston; and two, a line from Boston to Halifax, Nova Scotia, where ships from England and much of Northern Europe were accustomed to make their first landfall. Halifax was a hotbed of newshawks who used carrier pigeons and pony express relays to bring the latest news of Europe's wars, revolutions, and other major events to the big city newspapers. All the dailies in New York, for instance, were ready to pay good money for a front-page scoop on major news from the Old World.

In 1848, Morse agreed to a deal in which Smith traded his one-fourth patent rights in New England for an exclusive telegraph franchise in that region. This provided a larger "pot" of royalties for Morse and the two other patent-sharers, Vail and Gale. In addition, they were promised 3/16 of the stock in any companies Smith set up. Finally, Smith pledged a one-time payment of $37.50 for every mile of line that was built.

The early success of the New York, Albany & Buffalo (NYA&B) Telegraph Co., coupled with Smith's guaranteed access to Gotham, made the New England project attractive to investors. Within a year, the Boston-Halifax and Boston-New York lines were in service.

Whenever a ship hove into view, the Halifax office became a scene of indescribable chaos. Newshawks would dispatch their carrier pigeons to the deck of the ship, the captain would pocket whatever gratuity was in the message pouch, replace it with a few choice pages from the *London Times* or some other paper, and send each bird back to its owner. The chaotic scenes at the Halifax telegraph office eventually gave way, in 1850, to a more orderly arrangement, under the auspices of the new Associated Press (AP). The AP, in turn, began exerting pressure for special rates based on its high volume of often lengthy news dispatches. Smith balked at the large discounts that were demanded, and the response was a series of boycotts. The AP even resorted to its own telegraph, using a newly-patented (and inferior) non-Morse machine. The AP's point man in all this was a veteran Halifax news hawker named David H. Craig. Smith became convinced that Craig was behind the numerous incidents of sabotage that plagued the New England lines as the controversy dragged on.

The upshot of all this was that Smith's revenues proved to be disappointing. The problems were not helped by the severe New England winters, during which lines collapsed under the weight of icicles.

Early on, the $37.50 per miles payments to the patent holders fell into arrears. As the network expanded, so did the deficit. This added immensely to the ill-will that already existed between Smith and Morse. Faced with

Figure 3.1. "The Great Eastern in Hearts Content Harbor in 1866," Newfoundland. Western Union Telegraph Company Records, Archives Center, National Museum of American History, Smithsonian Institution.

litigation over the growing debt, Smith threatened to challenge the pending application for a patent extension, claiming to have proof that it was based on false documentation. The inventor's attorney shrugged this off with the comment, "He is the strangest mortal I ever met and this day I shall sue him."[3]

A private investigator was also hired to look into his private affairs, the intent being to find something that could be held over his head to assure future good behavior. Smith at this point was spending most of his time in Boston, home base of the New England telegraph. But hometown ties had recently been reinforced when he married a prominent Portland lady, seven years after his first wife, also a Portland native, had passed away. In Boston, the investigator took just days to discover that his quarry was supporting a mistress and two small sons. What Morse did with this intelligence is unclear. Smith did not make good on his threat about the patent extension, yet he would continue in his confrontational ways for years to come.

The inventor's effort to extend his patent had raised a storm of opposition from the various telegraph companies, all of which were eagerly looking forward to the day when they would no longer have to pay royalties. They filed suit in the D.C. District Court. On appearing for a hearing, Morse was surprised to see Smith sitting among the spectators. He wrote home, "FOJS is here, the same ugly, fiendlike dog in the manger he has ever been."[4]

Morse won his patent extension, but only after the challenge went to the Supreme Court. The vote there, for a seven year extension (1854–1861) was 6-3, and was based on the same hardship claim (numerous patent infringements) that had won an earlier extension. The effort to squeeze money out of Smith was not settled so expeditiously. The judge ordered mediation, and eventually a default judgment was issued.

At this point, in 1856, there was an unexpected breakthrough in the ongoing efforts to secure royalties from European users of the Morse telegraph. Emperor Napoleon III of France had convened an international conference to resolve the issue, and a number of countries, including the Austrian Empire (which stretched across nearly half of Europe) agreed to pay a "voluntary gratuity" to the inventor in the sum of 400,000 francs, payable in annual installments over a span of four years. With a one-fourth interest in the patent, Smith stood to collect 100,000 francs—or so he assumed. The inventor was quick to point out that this was not a typical business arrangement between patent holders and franchisees. The conference declaration was clear that this was a gift for the inventor.

Smith promptly filed a lawsuit. Both sides agreed to binding arbitration, and after lengthy deliberations, the decision came down in favor of Smith and the other patent partners, Vail and Gale, who hadn't asked for anything. The arbitration panel, however, assigned compensation to Morse for his years—long effort to secure European royalties. This would, of course, come out of the gratuity. The largest single item by far was a contract Morse had signed with a lobbyist requiring a one-third commission for any proceeds issued by the French government. Smith objected to this, noting that, as a minority patent holder, he should have been consulted in advance. He also made the point that not all of the gratuity was actually paid by the French government. (The man with the connections was a Belgian diplomat named Henrik van der Brock. France's share of the gratuity was actually 144,000 francs. Austria-Hungary and Russia kicked in 70,000 each. The Ottoman Empire gave 40,000; and the balance, 76,000 francs, was made up of contributions from the Papal States, the Archduchy of Tuscany, the Kingdom of Sardinia, Sweden, Netherlands, and Belgium.) The panel turned a deaf ear to Smith's objections. The net after expenses came to 247,000 francs. Smith's one-quarter share of that, translated into American money, was a little over $15,000.[5]

The notion of applying it to the $37.50 per mile debt he owed to his fellow patent holders never made it to square one. Smith's money troubles were clearly not limited to that one issue. In 1862, Judge Sprague of the New York Federal District Court issued an order to seize his property, on behalf of the patent holders, who hadn't received a cent on their per mile entitlement. Rather surprisingly, Smith had recently bought a house in Brooklyn—probably to be near the Lower Manhattan headquarters of the Western Union Telegraph Company, which was then buying up the telegraph fiefdoms, one after another. He was anticipating a big pay day, but wanted to secure a position in the company as well, presumably as its New England manager.

What he experienced now was a padlock and legal notice on his front door, and the disappearance of his horse and buggy from its stable around the corner. Not long afterward, Morse took it on himself to forgive the debt, after clearing the matter with his two lesser patent partners, Vail and Gale. He signed a release of all claims on Smith's property, but at that point half his furniture, plus the horse and buggy, had been sold. Smith then dropped his case against the lobbyist fee, and indeed signed an agreement whereby the two men renounced all claims against each other. This showed a remarkable magnanimity on Morse's part, a spirit of forgiveness that reflects well on his character.

Not long thereafter, Smith sold his telegraph empire to Western Union (WU) for the grand sum of $300,000 worth of WU stock. By then he was once again a single man, his second marriage having collapsed when Mrs. Smith received in the mail a family portrait from the mistress in Boston, accompanied by an explanatory note. The women in Smith's life received a substantial share of his fortune.

The $300,000 gave Western Union control of the busiest single line in the entire country—the Boston-New York route. But the New England system as a whole had a poor earnings record. Most years, Smith's main income was his salary as president, and the stock sold at a substantial discount compared with the New York, Albany & Buffalo Telegraph Co., which was by far the industry leader during these formative times.

Morse was generally content to leave the challenges of "inventing a system"—building an industry from the ground up—to businessmen and bankers. His salaried business manager, Amos Kendall, proved to be, on the whole, an honest and capable guardian of the inventor's interests. His one moral lapse was an involvement in lucrative, non-competitive construction firms that overcharged for materials and labor, and leaving Morse in the dark about them. Borrowing a page from the railroaders' playbook, Kendall built the Washington & New Orleans (W&NO) lines that way. He was also involved in at least one other such arrangement—in collaboration with "Fog" Smith and Eliphalet Case, on a line from Wheeling to Louisville.[6]

Morse did agree, however, to serve on the board of directors of the New York, Albany & Buffalo Telegraph Co., when it was founded in 1847 by some local businessmen in upstate Utica. He happened to have relatives there and was familiar with the town. Moreover, the patent royalty agreement granted him 700 shares out of an initial total of 3672, which made him the largest shareholder. Early on, there was an issue of high turnover among the telegraphers, and Morse suggested that higher pay might be the answer. The founding president, Henry O. Faxton, argued against this and got the board to vote the inventor's motion down. Morse then began missing meetings, and eventually resigned.[7]

Whatever Faxton's limitations in the realm of labor relations, though, it was clear that he was doing something right. The NYA&B proved to be the first telecom in history to turn a profit—and a healthy profit it was, with over $40,000 in dividends distributed to shareholders at the end of the first year of operation. The company's shares accordingly skyrocketed in value. This brought such a stark change in the inventor's customary penury that he immediately said goodbye to his long-time Greenwich Village attic and hello to a palatial estate on the banks of the Hudson River. There was no trouble securing a mortgage, which was collateralized by the inventor's NYA&B stock.

Faxton and his partners in an upstate delivery service knew nothing whatever about electricity, let alone the telegraph, when they incorporated their new company. He learned his new job, after a fashion, interning at the still unprofitable Magnetic Telegraph headquarters in Baltimore for a few months.

His relationship with the inventor took a sudden turn for the worse in January 1848. The cause was a nationwide newspaper advertising blitz that was launched by Henry O'Rielly against the Morse telegraph. In full-page ads full of bold print and exclamation points, the former Rochester newspaperman claimed that Morse was on the verge of being exposed as an impostor. The real inventor of the telegraph, these broadsides proclaimed, was a Bostonian named Dr. Charles Jackson. (This was the same Jackson who was Morse's fellow passenger aboard the *Sully*, returning from Europe in 1832.) The truth, moreover, was about to be revealed in a court of law—in a matter of weeks or months, if not days. For good measure, the ads also accused Morse of presiding over a monopoly that was gouging the customers.

There was indeed a court case in progress. It was a patent suit that Morse had filed against O'Rielly, relating to O'R's so-called Columbia Telegraph. The proceedings took place in Frankfort, KY, and the state court there was mulling over evidence that had been presented in a series of hearings that were held during 1847.[8]

Faxton informed the inventor that the company board was becoming uneasy, and suggested that the inventor might consider attending meetings again. In June 1848—with a decision still not forthcoming from the court— Faxton sent news that Morse's royalties would be held in an escrow account pending the court's decision. This set off a hot back-and-forth between Faxton and the inventor, who called the board's move "lawless" and threatened that "you will have trouble, depend on it." Faxton chided Morse on his "belligerence" and ignorance of the "needs" of doing business, and in one letter noted that "O'Rielly et al. have kept you at bay."

Morse's brother-in-law, Thomas R. Walker, happened to be a practicing attorney, and also happened to reside in Utica. Walker appeared at the company's offices several times, demanding payment. He decided to hold off any further action until the court announced its findings. When it finally did, in January 1849, Walker presented himself again at the NYA&B headquarters. Almost as soon as he came in the front door, he was handed a check.[9]

The NYA&B continued to prosper in the succeeding years. Relations between the two men, far from recovering, reached a new low when Faxton informed the inventor, in 1852, that he had been allowing "through traffic" from O'Rielly's competing system. After the Columbia telegraph was determined to be a Morse-knockoff and banned by the court in 1848, O'R adopted the new but inferior House Telegraph, which had just survived a patent challenge by Morse. Case told Smith the Columbia telegraph was "Morse with variations...so say intelligent men who have seen it in Cincinnati." He also called the House machine a "humbug." He explained that the contract with O'Rielly that Morse had repudiated back in 1845, and "which you would now be glad to recognize as valid," had never been declared null and void by any court.[10]

This dig was quite disingenuous. It overlooked the fact that the contract in question never had anything to do with operating a telegraph company. O'Rielly had been hired only to build telegraph lines for the Magnetic Telegraph Co., but had then taken it upon himself to found an entity called the Atlantic, Lakes & Mississippi Telegraph Co., without telling anybody. When confronted about it, he claimed that it was only a construction company. Morse then found fault with the lines O'R had already built halfway across Pennsylvania, and repudiated the contract on that basis.

The through traffic concession was actually based on a secret quid pro quo, in which O'R promised not to establish more than a token presence in the Empire State. It was generally believed that his system, despite its second-rate telegraph, shoddy line construction, and overall shoestring financing, was taking anywhere from 10 percent to 20 percent of the traffic away from the Morse companies.

The main reason was that telegrams sent on the O'Rielly system cost 50 percent less. The main benefit to O'R from this deal was that his Great Lakes business would now have access to Gotham via the reliable NYA&B lines, for a nominal fee.

Oddly enough, not long before Faxton astonished Morse with the through line revelation, O'Rielly had filed for personal bankruptcy. This did not affect the O'R companies, however. They would struggle on for years to come.[11]

The through line deal was not the only bad news the inventor received from the NYA&B chief. Morse was further informed that a letter he had once sent to Ezra Cornell (the construction superintendent of the original line between Washington and Baltimore and now the NYA&B's superintendent) had just been brought to Faxton's attention—presumably because it contained a reference to Faxton as "dishonest." As it happened, the latter had in his files a letter from Morse, confiding that Cornell "is not to be trusted." What conclusion was to be drawn from such slanders? Faxton had an answer: the inventor was, pure and simply, "a double-faced, Judas-like character."[12]

Not long after writing this, the NYA&B head moved on to other business challenges, at some remove from Utica. The fact that he had offered some prime Utica real estate in exchange for Morse's shares, during the months in 1848 when the inventor's royalties were being withheld, had become a source of embarrassment since Morse made sure that it became common knowledge. His successor, indeed, was none other than Attorney Walker. Years later he wrote to the inventor, professing a need to "get right with the world." He continued, "I hope to hear from you soon that I may make you ample restitution while my mind and means are sufficient to do so," and asked for an itemized bill. He never got one.[13]

The trans-Atlantic cable project made immediate headlines in the newspapers when it was announced in 1854. The widespread attention was inspired more by its boldness than by any measure of its feasibility. Only nine years earlier, the telegraph builders were pondering how to get a line across the Hudson River. The news caught Morse by surprise. Until then, his only concerns about Europe were about patent rights. These had been pursued to no effect at all until that very same year, when Napoleon III of France presided over an international conference that voted the 400,000 franc gratuity, which was granted in lieu of any future patent claims.

Equally bold was the financing. Investors who wanted an ownership stake had to put up $10,000 for each share. Morse signed on for one share, but the only person of his acquaintance who also made an investment was Taliaferro Preston Shaffner, the founding president of the New Orleans & Ohio Telegraph Co., which was headquartered in Louisville, KY. They had met during the O'Rielly proceedings before the federal district court in Frankfort, KY. The outcome, which debarred O'R from using the Morse machines under the "Columbia Telegraph" brand name, was a major relief for the fledgling NO&O enterprise, which was engaged in a head-to-head competition with the Rochester firecracker.

Shaffner came to Gotham and received his share directly from the hand of Cyrus Field, the impresario of the project. He also volunteered to drum up investor interest on the other side of the Atlantic. Field gave him carte blanche for this, provided he paid his own way and did not ask Field for commissions. Before embarking, Shaffner also visited Morse, and offered to seek a patent agreement in the two largest countries that did not sign onto the gratuity: England and Russia.

Russia was one of the few countries in Europe whose landscape was not already dotted with telegraph poles. Shaffner accordingly had two complete sets of Morse apparatuses with him when he arrived in St. Petersburg, and with the help of the American ambassador, Thomas H. Seymour, presented them to Czar Nicholas I. At that point, Russia was on the verge of going to war with its two erstwhile allies, England and France, over rival interests in the Balkans. Shaffner sold the Russians on the benefits of telegraphic communications between their capital and the fighting fronts, and was hired to build a line to the major Black Sea port of Sevastopol, 800 miles away in the Crimean Peninsula.

This was the largest single telegraph project ever undertaken, anywhere. Yet within a few months, with the help of rush shipments of wire and telegraph machines from manufacturers in Berlin, and labor provided by serfs who chopped down countless trees and planted the thousands of telegraph poles, a preliminary line was completed by the time hostilities began. The Crimean War, which lasted from 1854 to 1856, consisted in its entirety of an Anglo–French siege of Sevastopol. The telegraph to Sevastopol was never cut by the invaders; instead, it underwent improvements as the siege dragged on, and every Russian regiment maintained constant communication with the central command. (The invaders also had a local telegraph, and newspaper reporters used lines stretching away across Europe to make this the first war in history to be the subject of daily reports from the fighting front.)

One day (while the Crimean War was still in progress) the personal telegraph in Morse's study clicked out an urgent message from Cyrus Field. What did the inventor know, it asked, about a large crate of "telegraph equipment" that had just been offloaded on a Manhattan dock by a ship from England? It identified Field as the sender and Morse as the sendee. Without clearing customs, it was being reloaded aboard a ship bound for Germany when customs officers caught up with it. At that point the box had a new shipping label, with a "Colonel Shaffner" as the new sender, to a company located in Russia.

Customs agents soon appeared at Locust Grove to question the inventor. Shaffner, who was back in Louisville at this point, received a summons to appear before a military board in Washington. Morse did not hear from customs again, but he did hear from Shaffner, who wrote that he was going to spend "several years" abroad, and likely would never be seeing the inventor again.[14]

In fact, he turned up at Locust Grove three years later for a visit, with his family. It is unclear how he managed to wiggle out of his predicament with the mysterious shipment, especially since the crate proved to contain, not telegraph equipment, but artillery blasting caps manufactured in England.

When Ambassador Seymour inquired about this in St. Petersburg, he was told that the Russians thought "Colonel Shaffner" was a U.S. Army officer, acting with the approval of the American military. Shaffner, it turned out, actually was a colonel—a Kentucky colonel, with a certificate from the governor to prove it.

In 1856, just months after the Crimean War came to a rather indecisive conclusion, Morse took his wife, Sarah, on a trip to Europe. Thanks to Seymour, the Morses received an invitation to a reception at the Winter Palace in St. Petersburg. Next to them in the receiving line stood the former British prime minister, Robert Peel, and his wife. Czar Alexander II didn't have much to say to Peel, but he engaged Morse in some animated conversation about his marvelous invention.

Cyrus Field was a New York paper manufacturer of middling means when he was seized, in 1854, with the idea of building a trans-Atlantic telegraph cable. He signed a patent agreement with the inventor in February. In April, Field and the venerable (then sixty-three years old) Peter Cooper incorporated the New York, Newfoundland & London Telegraph Co. (NY,F&L)

The NY,F&L issued only 350 shares of stock at the extravagant price of $1,000 each, and stipulated a minimum purchase of ten shares. Morse subscribed for the minimum. Tal P. Shaffner borrowed money to buy his ten, while Field, Cooper and other Gotham investors bought a grand total of seventy more. The remaining 280 were sold to British investors, mainly through the personal efforts of Field and Shaffner, each of whom made separate trips to London, where the company established its headquarters.

Field and Cooper had no trouble convincing the inventor that such a high-cost, high-risk venture required the support of the entire industry. He gave them free use of the patent—no royalties—and the contract he signed even included a forgiveness of all through-traffic fees in the United States. (This required the consent of the individual companies, but Morse was prepared to make concessions on his personal share of royalties in order to win their approval.)

Amos Kendall, the inventor's business manager, had been completely left out of this deal. On being informed about it, he pointed out that Morse had no legal power to make this commitment on behalf of his patent partners and the various independent telegraph companies. Kendall also thought it extremely ungenerous that Field and Cooper didn't allow any discount at all on the ten shares they sold to the inventor.[15]

The only benefit for the inventor was his appointment as the company's "electrician," at a modest salary. In that capacity, he would supervise the preparations for and the actual laying of the cable. Morse was also made an "honorary" board member of the cable company, a position that included no voting right.

The inventor sought out whatever information he could obtain about the physical challenges that lay ahead. Michael Faraday, the noted British scientist, had recently presented a paper about a submarine cable at the Royal Academy. He sent Morse a copy. It contained the results of some tests on how *gutta-percha* (the standard wire coating) holds up under salt water. Faraday's paper also contained a proposed design for wire cables, from the coatings to the outer cover, and described how several wires should be deployed in a spiral pattern that would resist breakage. His plan called for huge master magnets, one on each side of the ocean.

While he was in Washington applying for his patent extension in 1854, Morse visited the Naval Observatory to inquire about the ocean depths and marine topography.[16] Unfortunately, there was no science of oceanography in those days. Coastal waters were well-charted to locate navigational hazards, but the deep sea floor was a total and complete mystery.

It wasn't until summer 1857 that all was in readiness for the laying of the cable. The project had been much anticipated by the newspapers, and now became the subject of daily news dispatches on both sides of the Atlantic. The American and British governments provided large Navy vessels—two from each country—for the line laying, at no cost to the company. The cable was manufactured in England, and wound onto huge rolls. Faraday and other advisers had determined that seven strands of thick gauge, coiled copper wire, surrounded by eighteen strands of coiled steel wire, and insulated by a thick coating of *gutta-percha*, would do the job. It was estimated that a line could transmit ten words per minute.

Morse spent the early months of 1857 in England, and then boarded the *U.S.S. Niagara*, Captain Hudson in command, after the great cable spools were loaded and in readiness. The *Susquehanna* (Captain Sands), *H.M.S. Agamemnon* (Captain Nodall) and *H.M.S. Leopard* (Captain Wainwright) were the other ships in the expedition.

In June the ships met in mid-ocean, cable was spliced, and they eased off—the Britishers heading for the Irish seacoast, and the Americans for Newfoundland. After three days the line broke, but was recovered and spliced. In the early hours of day twelve, however, with 250 miles of wire already sunk, the *Leopard*'s line broke with a loud crack, and was lost.

All four ships went back to England to replace the lost wire, and they started over. This time the project was completed without mishap. The Americans steamed into Trinity Bay, Newfoundland, and the British into Valentia Bay, Ireland, and on 4 August the lines (there were two—one for each direction) were declared ready for use.

On 18 August Queen Victoria sent the inaugural message, a greeting to President James Buchanan. After two weeks, however, the electrical impulses became noticeably and progressively weaker. After one month, the line was dead.

Faraday and some of his colleagues eventually found a likely cause. Glass, Elliott & Co., one of the cable manufacturers, had been storing it outdoors. The management admitted to noticing the *gutta-percha* melting in the sun, and after recoating some severely melted outer strands, had determined that the rolls were again fit for delivery, without unwinding to see if any inside wire was exposed.

Field was back in London in 1858, trying to drum up financing for another cable attempt. Morse was dropped as an honorary director of the cable company, and his position as electrician was approaching its expiration date with no offer of renewal. His irritation became greater when an attorney advised him that without any office in the company he was probably disqualified by law from serving in any capacity aboard a U.S. Navy ship.

Rather than verify this, he decided to resign his post as electrician forthwith, and avoid any confrontation.[17] (As matters turned out, the second effort did not involve any Navy ships.)

Field and Cooper were disappointed that Morse had not helped them secure free through-traffic on the lines of various telegraph companies. They had expected him to concede some of his patent royalties in order to win this concession, but the companies proved to be unimpressed with what was on offer.

The two had by then founded the American Telegraph Co (ATC). as an entity for obtaining access rights to the American lines. The name itself suggested that the trans-Atlantic project was no mere appendage of the American telegraph industry, but a wedge for gaining at least a measure of control over it. The business model that emerged, indeed, was no less ambitious than the effort to connect Europe and North America. It amounted to leasing existing companies, as many as possible, and consolidating the industry.

One would think that Field had enough on his hands with the cable project. He was obviously impelled by the sudden rise of Western Union, which was clearly bent on doing the same thing. WU was a kind of "wild card" (or "loose cannon") which, if left unchallenged, would probably end up with a monopoly on the American end of the trans-Atlantic cable.

The Magnetic Telegraph Co., which had not yet been approached by Western Union, was offered a twenty-five-year lease, with a guaranteed annual rental payment. This industry pioneer, still struggling to become consistently profitable, jumped at the opportunity. The stronger companies, such as the NY, A&B, and Kendall's Washington & New Orleans, drove a hard bargain. Kendall turned down an initial offer, even though he and Morse were fully convinced that industry consolidation was necessary. He finally agreed to a ten-year deal, provided the Magnetic was folded into it (thus cancelling its twenty-five-year commitment), and he became president of the Magnetic.[18]

Field approached O'Rielly when the NYA&B and the Southwestern Telegraph Co. proved obdurate. Up to his ears in debt (as always), O'R jumped at the offer. Royal E. House, whose telegraphs were the mainstay of the O'R network, then interposed himself. He felt slighted because the ATC did not speak to him directly about pending patent issues that he had with the Rochester hotshot. O'R had secured contract modifications that deferred his royalties, but then had missed a deadline for a $5,000 payment. In a letter to Field, House complained about the "most outrageous" treatment he was being subjected to, and threatened a lawsuit.[19]

The Civil War put a damper on the trans-Atlantic project. Relations with the U.K. were strained by several crises with the Union over neutral rights, and the W&NO, plus two-thirds of the Southwestern network, had been confiscated by the Confederacy. Border state telegraph lines and offices were favorite targets of Rebel saboteurs.

During this hiatus, Morse received many suggestions regarding the deployment of the submarine cable. Even his brother Sidney had an idea for a depth-monitoring "bathometer," and a buoy network that would suspend the cable at relatively shallow depths, and thus make any line breaks accessible for repair.[20]

The inventor, who owned one hundred shares in the NY, Newfoundland in 1857, increased his investment to six hundred shares in 1866, when the second try for a trans-Atlantic cable began. He did not participate at all in this new endeavor, and in fact was barely on speaking terms with Field, after learning of his dealings with O'R. All the wire for this second attempt (seven thousand tons of it) was carried on a single ship, the *Great Eastern*. The 680-ton behemoth steamer—the largest ship ever built—set out from Ireland in June, and reached Hearts Content, Newfoundland, on 27 July. Queen Victoria sent the first message, again, this time a greeting to President Andrew Johnson. It brought a prompt reply.

The NY, Newfoundland Company still had a predominantly British ownership, and its corporate offices remained in London. The minimum rate for a message was initially $100. This was reduced to $7.50 over the next five years, as two competitors—one a French company—spanned the ocean with cables. By the end of the century there were thirteen trans-Atlantic telegraph cables in operation.

A NOTE ON THE RIVAL TELEGRAPHS

Of all the alternative telegraphs that came onto the scene during the 1840s, only the House and the Wheatstone versions were in any way competitive with Morse. Wheatstone had a U.S. patent, but his system did not impress any American investors. The House device proved to be Morse's only seri-

Figure 3.2.   Telegraph key, 1844. Western Union Telegraph Company Records, Archives Center, National Museum of American History, Smithsonian Institution.

ous competition. When it was submitted for a patent, Morse was alerted by the Patent Office. He promptly filed a protest with Patent Commissioner Ellsworth (his former classmate at Yale), claiming patent infringement. The Patent Office required House to make some modifications, and then issued a patent in 1847. Morse still considered the House machine "a barefaced fraud."[21]

The distinctive feature of the House device was that it printed letters, rather than dots and dashes. House was a music teacher, and he noticed that telegraphers could decipher the Morse code by ear, from the pattern of sounds it made while clicking out its dot and dash messages. He thought that this "music" could be nuanced into "different tonalities" that could electronically activate the right keys on an alphabetical keyboard.

This involved replacing the binary dot and dash system with the twenty-four letters of the alphabet. While this was a marvelous improvement, it came at a great cost, both in money and intricacy. A Morse machine was a model of simplicity by comparison, and cost only $700 retail. House telegraphs rarely sold for under $1100. With their wiring and connections, they were more expensive to maintain. Telegraphers took longer to master them, and some simply couldn't.

Morse still wanted to file a patent suit, but his business manager, Amos Kendall, restrained him with the argument that the House telegraph could not compete, and so was not worth litigation. He kept Morse posted on O'Rielly's business, which made extensive use of the House telegraph. The reports were reassuring for a while, as O'R's market share seemed locked in at the 10 to 15 percent range, even though he charged 50 percent less to send a telegram.

An unexpected turn of events was Smith's lawsuit against the House patent, filed in Boston in 1851. It was prompted by O'Rielly's expansion into his New England territory. The O'R system relied on House telegraphs, but in this instance he was using them as backup for the newfangled Alexander Bain telegraph, which had no track record as a workaday machine. Smith's complaint was that both the House and Bain telegraphs infringed the Morse patent.

The case was dismissed, after the judge heard evidence in court, on the grounds that Smith offered no compelling evidence to prove that the Patent Office had been wrong in granting patents to House and Bain. (Hugh Downing was president of O'Rielly's New England Telegraph.)

Despite this, Morse finally pushed ahead with his own action against House in 1854. The fact that House telegraphs were still in use had made Kendall's advice—that House wasn't to be taken seriously—less convincing. What got Morse moving was Kendall's latest report on the O'Rielly's companies: they had collectively netted $60,000 in profits the previous year—a sharp rise from the $40,000 of the previous year.

Morse began to fear a bandwagon effect from this news as Cyrus Field, the trans-Atlantic pioneer, obtained House patent rights for his new American Telegraph Co. ATC already had a license for the Morse patent. The inventor inquired about this, and Field replied, unconvincingly, that he was just "neutralizing" the House patent—tucking it away, so to speak, to keep it out of reach of any competitors on the trans-Atlantic route. His letter concluded with a suggestion that the House and Morse patents ought to be "consolidated."[22]

Morse v. House (1854) was argued in the NY Federal District Court. Noting that the Patent Office had already acted on all of Morse's objections before issuing a patent to House, the judge wondered what might entitle him to overrule the technical expertise that produced that decision. Like Smith's lawsuit, the case was dismissed.

As the years went by, concern over the House telegraph waned. James D. Reid wrote Morse that it was "too complex, noisy and expensive."[23] Anson Stager, who had succeeded Cornell as president of the Lake Erie Telegraph and was now a Western Union vice president, provided Morse with a detailed assessment of the shortcomings of the House machine, based on his earlier experience with it as an O'Rielly manager.[24] Several Western Union offices used House machines, mainly as testing sites for improvements that were tried from time to time. But the Morse telegraph was the mainstay at WU, as it was everywhere else in the world.

ALEXANDER BAIN'S TELEGRAPH

Morse and Dr. Jackson had discussed, aboard the *Sully*, two ways of recording date on a telegraph. One was to use an electrical stylus or pen, and the other was a pattern of spark-induced burns on chemically treated paper. The inventor started out favoring the latter, but he ended up with a marking pen. Bain's telegraph, which was patented in England in 1846 and the United States in 1848, opted for the chemical approach. Bain, a Scotsman, called it a "chemical telegraph."

The Patent Office actually rejected Bain's application, noting that aside from the chemical recorder, it resembled the Morse telegraph—including its dependence on the Morse code. The Scotsman accused the patent officials of inconsistency. If they could approve the House machine, which also had such strong resemblances to Morse's (even without the Morse code), he felt they should approve his. As for the Morse code, which he assumed to be the major sticking point: it was not Morse's invention. He demanded a review.

His claim regarding the Morse code was based on a permanent exhibit at the American Museum in Philadelphia. On display was an ingenious messaging system that had been developed by a Philadelphian named Thomas Swain. It was a way of communicating through the walls separating adjacent rooms. Swain used two symbols—a knock and a scratch—to formulate an alphabetical code that bore a striking resemblance to the dot-dash Morse code. Swain called his system "mural diography," and it predated the Morse code by several years. Had Morse or Vail ever walked into the American Museum, the essentials of the Morse code were right there waiting for them.

Bain was turned down again, and took his case to the federal district court in Washington. The Court overruled the Patent Office and awarded him his patent. Bain then made a curious move: he arranged a private meeting with Morse. They debated the similarities between the Morse and Swain codes, and then Bain offered to sell his patent to Morse. The offer was rejected, with a promise that the district court decision would be appealed. (A note summarizing the meeting can be found in the Morse Letter File for April 1848.)

Oddly enough, Bain decided to backtrack a little. He abandoned the dot-dash Morse code paper markings in favor of a disconnected zigzag-like pattern—a zig and zag being the equivalent of a dot and dash.[25] This put greater demands on his dicey "chemical telegraph."

Bain secured some British financing for a telegraph from Washington to Baltimore. It paralleled the original Morse line that now belonged to the Magnetic Telegraph Co. In November 1849, the Magnetic line filed an appeal of the Washington District Court's reversal of the Patent Office. The case was thrown out when the judge decided the company lacked legal standing in a patent case.

In April 1851, Morse himself filed an appeal. His main objective was to rescue the Morse code as his own original idea. The fact that O'Rielly's appeal of the 1849 Frankfort Decision was pending before the Supreme Court lent this action an air of urgency, even though it had never occurred to O'Rielly to question the Morse code, specifically, in his case against the Morse patent.

Morse v. Bain was scheduled for trial in Philadelphia. Before the hearing date, Bain's Washington-Baltimore line began winding up its business, and O'R informed Bain that his equipment in NY and Boston was being replaced with House telegraphs.

The case ended rather suddenly with a consent agreement, and the court hearing was cancelled. The Morse code issue was not resolved, but neither did it generate any more publicity. Bain went home to Scotland, never to return.

THE WESTERN UNION TELEGRAPH COMPANY

Hiram Sibley was a merchant in Rochester, NY, who, like everybody else in town, was mightily impressed with the exploits of native son Henry O'Rielly. These exploits were reported almost daily, year in and year out, in the *Rochester Argus*, the newspaper O'Rielly founded and ran before moving on to Gotham. The Sibley-led "Rochester Group" (so-called by Amos Kendall) incorporated the New York and Mississippi Valley Telegraph Co (NY&MV) in 1851, and obtained a license to use the Morse telegraph. This obviated any role for O'Rielly, the arch-enemy of the Morse interests who was struggling along with his House telegraphs. The NY&MV was not a typical telegraph company. It was intended as a vehicle for consolidation, a prototypical holding company that would exercise a centralized control over other telegraph companies.

The NY&MV started out by leasing (rather than buying) "side lines," the lines that branched off the big city connectors to reach the smaller communities. Money for buyouts was not available, and even the leases were highly dependent on lines of credit from several banks in New York State. From the outset, however, the Rochester men regarded a lease as only a stepping-stone to outright ownership. In this it anticipated the American Telegraph Co., which was formed a few years later and had access to credit from the large Manhattan banks.

In 1853, the Group helped to organize an industry-wide gathering, for the purpose of resolving territorial disputes and coordinating rates and through traffic policy. The result was the American Telegraph Convention, and the American Telegraph Federation. Many of the companies viewed the Federation as a bulwark against takeovers by consolidation interests, which might pursue their ends via rate wars. The Rochesterites, however, were not inclined to hostile takeovers. They viewed the Federation as a stepping stone to consolidation that would be based on mutual interests.

The Rochester Group's "breakout" year was 1857, when the NY&MV signed ten year lease deals with Cornell's Erie & Michigan, the New York, Albany & Buffalo, the Southwestern, and the California Telegraph Co. (CTC) (which was headquartered in San Francisco).

The California line was the only regional telegraph service west of Omaha, NE. It was the Group's bold vision not only to consolidate the industry, but to establish a coast-to-coast network. Ezra Cornell came up with a new name for the company, Western Union, which aptly summarized its mission.

The lessees got seats on the company boards, but they always reassured their targeted companies that they were not going to push for management changes, refinancing, reorganization, or anything that would dilute existing stockholders' equity. Added to these reassurances were grants of NY&MV stock to the directors and senior management of the leased companies.

The American Telegraph Co. (the creation of Field and Cooper) was still very much alive in 1857. Although it was now clearly the underdog in the consolidation race, there was a distinct possibility that the industry might become divided into two warring camps, with rate wars and other indignities. The two sides got together and worked out an agreement called the Treaty of the Six Nations—a whimsical reference to the legendary Iroquois Federation of six Indian tribes in colonial New York.

There were in fact six parties to the "treaty," and all overlapped the former stomping grounds of the late Iroquois empire. The Six were Field's American Telegraph Co., WU, the three Morse companies (including NYA&B) that WU had just leased, and the still-independent Illinois & Mississippi Telegraph Co., a Chicago company that (like the others) used the Morse machines.

In 1856, American signed a twenty-five-year lease with the Magnetic Telegraph Co.—Morse's original line, between Washington and Baltimore (and now extended to Philadelphia). It then turned to the Washington & New Orleans. Amos Kendall agreed to a ten-year lease, provided that the ever-troubled Magnetic was folded into it (thus cancelling the twenty-five-year agreement). In 1859, American bought the merged Washington & New Orleans/ Magnetic Telegraph Co. outright. Kendall, Morse, and the other shareholders were paid mainly in American stock. Kendall remained as chief executive. That same year, American struck a lease agreement with Smith, for his New England system. "Fog" had earlier turned down an offer from WU, which wanted to dilute his powers as president, due to a perceived managerial weakness. Under American, his authority was untouched.

Overall, ATC's collection of leased companies was notably less profitable than WU's. The O'R network was an especially serious drag on revenues. Only Kendall's old Washington & New Orleans lines earned a consistent and healthy income. But when the Confederacy confiscated the lines in 1861, ATC's ability to compete with Western Union in the consolidation race was severely compromised.

American's only hope was to outbid Western Union for the Federal contract to build a transcontinental telegraph. The Pacific Telegraph Act of 1860 offered an annual $40,000 subsidy, but stipulated that the charge for a ten-word transcontinental message had to be $3 or less—preferably less. This was a tall order for a service that would include eight hundred miles of essentially Indian country, where the customers were few but the hazards many. Reliable rumors surfaced that WU was readying a $2.50 bid. ATC withdrew from the competition.

Western Union had a head start via its 1857 lease of the California Telegraph Co. CTC was based in San Francisco and operated as far east as Carson City, NV. A thousand miles of Indian country separated the NV capital from the telegraph offices in Omaha, NE, and St. Joseph, MO, which were both connected to St. Louis.

Western Union built a line across that thousand mile gap in just one year, after winning the Federal contract by default. In 1861, the inaugural telegram, originating on Telegraph Hill in San Francisco, reached President Lincoln at the War Department, across the street from the White House.

Morse never met the Rochester people, and did not venture to invest in WU until 1863, when he traded $200 worth of his New York, Albany & Buffalo shares for WU stock. Thomas R. Walker was president of the NYA& B at that point. As an attorney in Utica, NY, where the company offices were, Walker had once confronted his predecessor, Henry Faxton, over Morse's withheld royalties. That was back in 1848.

In 1864, Morse traded all 950 of his NYA&B shares for Western Union stock, upon hearing from his brother-in-law that WU was about to buy the company outright.[26] That same year, Cyrus Field threw in the sponge on his consolidation scheme. His entire American Telegraph system, including the O'Rielly and Smith leases, were sold to Western Union.

In 1867, Western Union completed the consolidation of the telegraph industry. All the stock Morse held in those companies (mainly via patent royalties) became WU stock—7490 shares in his own name, and nearly three thousand more in his wife's, children's, and other relatives' names in all.[27]

The inventor knew the hazards of having "all my eggs are in one basket," but he was relieved that the dog-eat-dog days seemed to be over for the telegraph industry.[28] Several companies that had resisted WU's terms joined together in the U.S. Telegraph Co. (UST) in 1864, but UST finally succumbed in 1867. The telecommunications industry was now run by a monopoly—just what Morse had always believed to be the only viable business model for his great invention.

Western Union paid a regular 2 percent quarterly dividend, based on par value of $100 per share. (Payouts were $2 per share every three months.) Only twice—in January 1866 and January 1867—was the dividend ever late. Both times, the awareness of having "all my eggs in one basket" caused the inventor acute anxiety.[29] But he never sold a single share.

The company built a new headquarters building at 125 Broadway in 1866, about two miles from the inventor's winter home on W. 23rd Street. Some executives lived in his neighborhood. But Morse visited the Broadway office only a few times. President Sibley, et al., would turn up at the inventor's Tuesday evening soiree on 23rd Street just once a year, at Christmastime.

Sibley retired as the founding president in 1865, after losing $5 million on an ill-fated project for a telegraph to Europe by way of Alaska and Siberia. He was succeeded by Jeptha Wade, the erstwhile vice president for Acquisitions. Wade was a former O'Rielly man. The year he took over, the company had more than five thousand employees, and it operated seventy-five thousand miles of telegraph lines.

By the time Morse died in 1872, Wade had also retired and been replaced by William Orton, who was formerly O'R's top lieutenant. The number of employees then was over eight thousand, and the average pay of a telegrapher was $75 a month, up from $60 in 1865. The company's lines extended over 137 thousand miles. The number of Morse telegraphs in use increased from three thousand in 1865 to more than seven thousand in 1872. The number of messages sent increased from six million per year to twelve million during that interval.

It is a remarkable fact that every new technology requires the invention of a means of delivery to its market. Morse was aware that any delivery system for the telegraph could only be as strong as its weakest link, and this fact was borne out by experience during the nascent industry's "feudal" period, from 1844 to about 1864. It had to go through that period, however, because the invention was so new fangled that people were hesitant to invest the large sums that were required to build and maintain a centralized system. Growing confidence in the viability of Morse's invention, via the success of some of the regional networks, made consolidation inevitable.

The inventor himself played a largely passive role in the process. This is not unusual, since the process is mainly the province of businessmen. The challenge for them was actually rather daunting, considering that the telegraph was a totally unprecedented technology.

Their task was eased somewhat by the simplicity of the Morse machine. A more complicated device (such as the House or Wheatstone telegraphs) would have made them more dependent on expert technicians, thus increasing operating costs. Maintenance and repair costs would have been higher, and customer satisfaction lower. Bell's telephone, grafted onto such devices, would have suffered from the same handicaps, and so indeed might the later advances in telecommunications...until somebody came up with the basic Morse model—digital and streamlined, the image of simplicity.

The role that Henry O'Rielly and his Rochester acolytes played is remarkable. They were centrally involved at every stage of the industry's development, from the very beginnings through the "feudal" decades and on to that great unregulated monopoly, Western Union. After Morse, these heretofore forgotten men were the true founders of the telecom industry in the United States.

NOTES

1. Henry O'Reilly Papers. New-York Historical Society. Newspaper clipping, 1845 Folder.
2. Ref. F. O. J. Smith letter in New York Herald, 15 April 1850; The State v. Smith. Brown, Thurston & Co., Portland, ME, 1866, p. 11, 26; Francis O. J. Smith Correspondence (microform), New York Public Library, Case—Smith, 25 April 1848; Case—Smith, 31 August 1847, 12 November 1847, 12 December 1847; 20 March 1848 (here Case calls O'Rielly, "your devil"), 12 April 1848, 25 April 1848, 2 May 1848, 29 July 1848; 5 October 1848; Case—J. J. Speed (partner of Ezra Cornell in a Lake Erie region telegraph venture), 25 April 1847.
3. E. Fitch Smith—Morse, 2 April 1850.
4. M—Sarah Morse (wife), 13 April 1854.
5. M—Kendall, 3 August 60; M—Smith, 20 November 60, 15 May 1862.
6. Case—Smith, 25 April 48, (Smith Corres., NYPL).
7. Faxton—M, 18 March 47.
8. Ref. Morse Letter File 1848; Faxton—M, 9 March 1848. The Journal of Commerce (New York) editorialized in favor of O'Rielly's charges of price gouging. Morse's brother, Richard, riposted with an angry letter to the editor, and cancelled his subscription. (Richard Morse was himself a prominent New York newspaperman.) A detailed account of the O'Rielly patent trial can be found in Alfred Vail.
9. Faxton—M, 15 March 1848; M—Daniel Lord (Morse's attorney), 13 November 1848; Faxton—M, 17 November 1848; M—Faxton, 26 November 1848; M—Lord, 29 November 1848.
10. Faxton—M, 17 August 1852.
11. Answer of Donald Mann as Trustee for Henry O'Rielly, Superior Court, City of New York, to J. Oakley, Receiver. It states, "Henry O'Rielly should assign to said Receiver all his property, real and personal, with certain exceptions." The document was signed by O'R and dated 19 March 1851. It indicated, among O'R's holdings, "heavily mortgaged" acreage in Virginia and Indiana. Presumably this was right-of-way for unbuilt telegraph lines.
12. Faxton—M, 18 June 1852.
13. Faxton—M, 11 January 1862.
14. M—Shaffner, 8 April 56; Shaffner—M, 28 April 56.
15. Cf. Faxton—M, 11 Feb 1857; Kendall—M, 4 Jan 1857, in which Kendall refers to the Field group as "hostile interests."

16. Lieutenant M. F. Maury—M. 23 February 1854.
17. M—Sidney Morse (brother). 15 March 1858.
18. Kendall—M, 22 May, 18 July 1856.
19. House—Field, 8 December 57. Cyrus Field Archive, New York Public Library.
20. M—Field, 22 June 1866 (Field Archive).
21. M—Smith, 8 May 1847.
22. Field—M, 18 Apr 1854 (Morse Letter File).
23. James Reid—M, 14 Aug 1860.
24. Western Union Telegraph Co.—M, 22 August 1860.
25. Kendall—M, 13 April 1849.
26. M—Edward W. Chapman (Treasurer of the NYA&B), 2 January 1864.
27. M—Richard Morse (brother), 8 December 1866.
28. M—E. S. Sanford, 12 December 1866.
29. Ibid., 12 December 1866.

Chapter Four

The Question of Origins and Originality

Did Morse Really Invent the Telegraph?

It took Morse five years, from 1832 to 1837, to devise a functioning telegraph. During that entire period, he showed no awareness (none, at least, in his correspondence) that he was in a race with others. Had he learned, for instance, that a Russian scientific dilettante named Baron Paul Schilling also began working on a telegraph in 1832, and had a working model by 1835, the news would have come as a shock. Morse was also blissfully ignorant of similar endeavors by others—Charles Wheatstone, William Fothergill Cooke and Sir Humphry Davy, in England, and Carl August von Steinheil, in Munich, Bavaria.

This sudden efflorescence of telegraphic research in Europe had a common origin. First and foremost was the publication, in 1820, of Dominique Arago's pathfinding paper on electromagnetism. Second were the promising experiments in electrical conductivity and magnet design, especially those of Michael Faraday in England and Marcel Sturgeon in France, that were inspired at least in part by Arago's research.

If he had any vague suspicion that the scientific advances were inspiring anybody besides himself—a suspicion that would have been very reasonable—Morse never expressed any concerns about it. The fact that his endeavors were single-minded and unremitting for five years, as if he were in a race with the clock, might suggest otherwise. But an awareness of any of these rival efforts would immediately have raised the question of whether they were better than his own. Morse never made such an inquiry (which would surely have occasioned a suspension of his own project until he found out).

The competition, meanwhile, were at a serious disadvantage: none had even the slightest awareness of this outlier in his Washington Square loft, whose concept was so obviously superior.

In the end, people were deeply suspicious about this portrait painter-turned-electronics wizard. The suspicion was especially strong among the investors and business promoters who wanted to make money off the telegraph, but were not so keen about paying royalties to the inventor. They were even less keen about the prospect of being brought up short by one of the several rival claimants to the invention. Some hired expensive lawyers and detectives to sniff out the truth.

Inevitably, there were litigious debates, in various courts, including the U.S. Supreme Court, about the difference between science and technology; the relative importance of concept and implementation in the process of invention; and finally, what qualified somebody to be considered a co-inventor, and not just an assistant to the inventor. More than any other invention in American history, the Morse Telegraph brought some definition to these important questions. The controversy would cast a long shadow over all future patent litigation.

The question posed in the title of this chapter can best be answered by examining the activities and recollections of the people who had any connection with Morse's project between 1832 and 1837. There are just five men—Professor Leonard Gale, Alfred Vail, Professor Joseph Henry, Harrison Gray Dyar, and Dr. Charles T. Jackson—who fall into this category. Their respective stories will give us the clearest picture possible of what Morse actually did. Collectively, they comprise the only eyewitness account of this epochal event—the invention of modern telecommunications.

PROFESSOR LEONARD GALE

Professor Leonard Gale was awarded a 1/16th share of the Morse patent for "effecting improvements in the philosophy and physics" of the telegraph.[1] He was professor of geology and mineralogy at NYU, and like Morse, a member of the university's charter faculty. Gale first laid eyes on the telegraph in January 1836. He thereafter became a regular visitor to the Morse art studio, which was on the floor directly above his classroom and laboratory. With some help from an NYU science colleague, Professor John Draper, Gale improved the design of Morse's all-important combined circuit (relay) system, giving it more power. He stated in sworn testimony during patent suits, however, that the combined circuit relay was Morse's alone, and was notably different from the one invented by his friend and colleague, Joseph Henry, in 1835.[2] "Fog" Smith did not believe it, and neither did Henry (who never submitted his relay for a patent).

When Morse left NYU for good in autumn 1843 to build the Washington-Baltimore line, Gale stayed on, teaching chemistry. He visited the inventor in Washington occasionally, offering technical advice for a small stipend from the construction budget. Later on, he made visits to examine the Magnetic Co. construction northward from Baltimore. In 1845, he and Vail inspected the workmanship on a Philadelphia-to-Pittsburgh line that was nearly half finished under a contract with Henry O'Rielly. Gale wrote the report that resulted in the repudiation of O'R's contract. That report marked the beginning of a long history of legal feuding between Morse and the intrepid entrepreneur from Rochester, NY.

In 1847, Gale left NYU for a much better paying job as an examiner at the Patent Office in Washington. Commissioner Ellsworth—the former Morse classmate who issued the telegraph patent in 1840—left shortly after hiring Gale, in order to run for a seat in Congress from his hometown, Hartford, CT.

During Gale's early years in Washington, three rival telegraphs (the House, Bain, and Hughes systems) were submitted for approval. Gale recused himself from these applications, but he was able to keep Morse abreast of every detail of the examiners' assessments as to their originality, strengths, and weaknesses. The advantages of having an insider at the Patent Office, however, were more apparent than real. Morse sued House and Bain for patent infringement, and eventually lost both cases.

Gale's presence in the Patent Office had a more telling effect when the inventor applied for a patent extension in 1854. This was hotly contested by various parties who wanted to operate telegraph companies without having to pay royalties. When the patent was renewed, opponents filed a lawsuit, and when this failed in the D.C. District Court, they appealed all the way to the Supreme Court. Through it all, Gale was able to keep his partner up-to-date on the issues, and especially the stance the patent commissioner would take when called upon to testify. For his clandestine services in these proceedings, Morse paid Gale $1,000.[3]

In July 1850 the inventor bought Gale's 1/16 patent share, paying for it with a $15,000 6-percent interest-bearing note. This was even larger than the mortgage Morse took out for his country estate, Locust Grove, in 1847, and payments over the first five years were so erratic that they would have strained a less solid friendship. Gale never complained, and agreed to a refinancing proposal in 1850. The loan was finally paid off in 1859.

In 1857, Gale was suddenly fired from his job. In bitter letters to Morse, he blamed it on patronage politics.[4] He tried his hand as a "patent consultant" in Washington for a while, but there was little demand for his services.

Gale then found a position at a chemical company in Brooklyn, NY. He and Morse met on rare occasion, usually at some public event. Their last meeting took place at the unveiling of a bust of Morse in Central Park, in 1871.

Chapter 4
ALFRED VAIL

Vail's role in the creation of the telegraph was described in the second chapter. From the time he joined Morse, in 1836, to the time the patent was granted, in 1840, the device underwent crucial transformations that finally made it viable. Since the two men worked practically elbow-to-elbow, this has led to much speculation as to Vail's role in all this. In 1836, for instance, the inventor was struggling with his "portrule," an unwieldy printing device that contained all the letters of the alphabet in Morse code. The swing-like mechanism engaged with a moving roll of paper, and the synchronization of the two moving objects was a challenge that finally could not be overcome. The portrule, therefore, was not a feature of the 1840 telegraph. And as for the Morse code: Chapter 2 notes that Vail did the research on the alphabet. The idea was that the most frequently used letters would have the shortest code, and the longer combinations of dots and dashes would be reserved for the least used letters. So, in accordance with the inventor's instructions, Vail "wrote" the Morse code. (This should not be confused with the invention of the code, as a binary system of dots and dashes.)

So we know that he had something to do with abandoning the portrule, and was therefore in on a major redesign, and had something to do with the final form of the Morse code. The portrule gave way, ultimately, to the telegraph key. Vail was there when it happened. Was it his idea?

Vail never claimed any more credit than Morse granted him, which was certainly not negligible. His archive at the Smithsonian Institution contains no such claims, either, although a letter from his brother George advises, rather cryptically, that he should "keep quiet" about his contributions to the telegraph and not challenge Morse "in any way."[5]

Vail was assistant superintendent of the experimental line in 1844, and held the same position in the Magnetic Telegraph Co. when it was formed in 1845. His $1,500 salary on the experimental line was continued by the Magnetic. It lifted Vail and his young family out of a condition of near-penury, but low revenues on the new company caused his pay to be cut in half in 1846. Vail then became Superintendent at Amos Kendall's new Washington & New Orleans Telegraph Co., at a salary of $1,000.

Vail's requests for a raise during his second and third years at the W&NO were turned down, and the young man's relationship with Kendall deteriorated. He advised Morse that he didn't trust Kendall, and that the messy files at the W&NO hid evidence of financial legerdemain.[6] Kendall had a steady correspondence with the inventor, as his personal business manager, but never complained about Vail.

In 1845 Vail wrote a book about the new invention, titled, *The American Electro-Magnetic Telegraph* (Philadelphia: Lea & Blanchard). This was followed in 1847 by an updated and expanded version, *The Early History of the*

Electro-Magnetic Telegraph. The latter is the most authoritative and detailed account of the invention of the Morse telegraph, and gives a detailed description of the original transmitter and receiver. It also contains a thorough account of the various forerunners and rivals of the Morse telegraph. Its technical detail even qualifies it as a primer on electrical engineering—perhaps the first one ever written.

Vail included in his writings the scientific underpinnings—the research in electro-magnetism—that underlay the telegraph. For some reason he downplayed the very important work of Joseph Henry, the Princeton professor whose contributions to the study of electro-magnetism were widely recognized in the scientific community. Henry was especially offended by the assertion (first made in the 1845 edition) that Morse turned his rudimentary relay (the combined circuit that allowed an electrical signal to travel indefinitely) into a workable product. (Morse patented his version of the relay in 1846, in a revision to his original 1840 patent. Henry, whose first relay was successfully tested in 1832, never applied for a patent.)

Henry was by no means a publicity seeker, but the issue even found its way into the newspapers. Morse stood by his assistant's characterization, and did not try to hedge or revise it in any way.

Vail quit the W&NO in 1852, shortly after his wife passed away. He asked Morse to design a gravestone inscription for her, and in the same letter wrote, "I have made up my mind to leave the telegraph to take care of itself, since it cannot take care of me."[7]

He remarried a year later. By then the stock dividends he received as a 1/8 patent holder were bringing in more than pocket change. He worked with his father and brother at the family business, Speedwell Iron Works, in NJ, and then found a job with the Baldwin Locomotive Co. in Philadelphia.

In 1856, Vail loaned most of the earnings he had made from his patent royalties to a New York hardware manufacturer, at a generous 8 percent interest. The company went broke, and Vail lost all his money. Health problems then began causing him to miss work. With a wife and three children to support, he began selling the telegraph stock he had received as a royalty. (Some licensees were allowed to pay in stock rather than cash, or sometimes a combination of both.)

Morse was protective of his former right-hand man. When advised by Kendall, for instance, that Vail had no claim on the 1846 patent revision, which mainly covered the relay, Morse brushed him off, saying, "This is news to me," and made sure Vail got his usual 1/8 share.[8] When the Supreme Court approved the patent extension in 1854, the decision specifically said the approval was for the inventor alone, "not to Assignees"—meaning, in the main, companies that paid for patent rights. Morse considered using this as a way to shake off Smith, who was an "assignee" of a sort, but this would have entailed shaking off Vail, for whom the seven-year extension was a pure windfall.

Vail's patent share was actually 1/16, because he had given half to his brother George. Morse knew this, and in fact corresponded with George occasionally, during the years when he was managing production of the early, prototypical telegraphs.[9] In later years, though, George took up the cudgels against the Magnetic company, after it cut his brother's pay in half. He wrote Morse, comparing Vail's pay cut with the president's $2,000 salary, and calling the company's board of directors (which included Kendall) a gang of "bloodsuckers." Then, when Alfred couldn't obtain a raise at the W&NO, George complained to Morse that Kendall was "opportunistic."[10]

In 1854, he also submitted a deposition on behalf of Morse's patent extension request. It carried some weight, because he had built model telegraphs at Speedwell during its formative years, and also happened to be a member (since 1852) of the U.S. House of Representatives. (When Congress was not in session, he continued working at Speedwell, and took over the business when his father, Stephen Vail, retired.)

Alfred Vail died in 1859, at the age of 52. His estate consisted of little more than the 1/16 royalties that would end when the extended 1840 patent expired in two more years. (The 1846 revised patent had not been extended, and was just one year short of its expiration date.)

Then the European Gratuity materialized, with the first of the four installments arriving later in 1859. Vail's widow was living in Speedwell at that point, largely dependent on her in-laws for support. The Articles of Agreement contained no provision for survivor benefits, but Morse wrote out a personal check for Vail's one-eighth and journeyed to central New Jersey in order to hand it to Mrs. Vail.

In 1861, Morse secured a job at the American Telegraph Co. for Vail's son, Stephen. The following year, he steered the young man into a higher-paying job at the North American Life Insurance Co., which had given the inventor stock and a seat on its board in return for the privilege of using his name in its advertising.[11]

In 1862, Morse made another trip to Speedwell. On that visit, Mrs. Vail signed a general release from any future obligation the inventor might have to Vail's estate under the original Articles of Agreement. He then gave her a check for $5,000. It was the last installment of the European Gratuity.

Two years later, the widow Vail wrote to Morse asking if he could dispose of her last shares of telegraph stock for a good price.[12] He bought them himself.

William Vail was then preparing a new edition of his father's telegraph history. Morse agreed to check the galleys and made some corrections—the year of the patent caveat to 1837 from 1838, etc. Not changed at all was the brushoff of Joseph Henry.[13]

Vail had willed a collection of early telegraph artifacts to Morse. These were stored at Speedwell until 1870, when Vail's second son, J. Cummings Vail, sent them to the inventor.

The following year (the year the statue went up in Central Park) Morse's birthday was the occasion for a black-tie gala at the Academy of Music. After various speeches and toasts from Gotham's elite, Morse sat down before a telegraph onstage and tapped out a message. It was a greeting to the new office in Hong Kong. Less than half an hour later, as more encomiums were offered to calm the growing suspense, a reply came back. The crowd burst into applause. When the noise died down, the inventor walked to the front of the stage. In an even voice he recalled the humble beginnings of his great adventure, and how lonely and frustrating it had been, until a member of his small class of art students, Alfred Vail, joined in his daily endeavors. He regretted that Vail was not present and standing there on the stage, to share in the accolades of the assembled throng. Cummings Vail read this tribute in the newspaper the next day, and sent a note of thanks.[14]

After Morse's death, in 1872, Mrs. Vail began claiming that the principal rewards of the invention rightfully belonged to her late husband and his heirs. E. Cummings Vail published a new edition of Vail's *Early History of the Telegraph*. It contained an introduction that claimed his father to be the "co-inventor" of the telegraph and the Morse code.

Perhaps Vail confided these sentiments to his wife, years before. But perhaps not. Certainly, the family fortunes would have been better if the Speedwell Iron Works, which built the first telegraphs, had endeavored to become a major supplier of the instruments. Aside from taking advantage of their insider status, the expert oversight of Vail and his brother George would surely have led to improvements that they could have patented.

Instead, the main manufacturer of Morse telegraphs was the Proscher Co., of Manhattan—later to be replaced by the Charles G. Page Co., also of Manhattan. Siemens-Halske was the main manufacturer in Europe.

JOSEPH HENRY

In 1837, with his first telegraphic device nearing completion, Morse was told by one of his NYU colleagues that Henry had written a brief essay in 1831 titled, "Barlow's Project on the Electro-Magnetic Telegraph." It gave the inventor "a sleepless night." That somebody else had conceived the idea of a telegraph—and an electro-magnetic one at that—was quite a rude shock. Before long, he satisfied himself that Barlow's project had never amounted to anything. The experience inured him to discoveries yet to come of sundry telegraphic enterprises, so numerous that they might have overwhelmed his resolve had he learned of them years earlier.[15]

The inventor's acquaintance with Henry's experiments in electro-magnetism dated from 1836.[16] These had nothing to do, as far as he knew, with inventing a telegraph or any other machine. Henry actually "invented" an electric bell ringer in 1831, but it was intended only as a laboratory tool—using the loudness of the bell to measure the strength of electrical impulses along a wire. His project had been inspired by an article in *Silliman's Journal of Science* in 1831 (Vol. 19) that described a new magnet designed by the French scientist, Sturgeon. Henry perceived a way to strengthen the magnet, by winding the wire more tightly around the iron horseshoe, and insulating the wire against physical contact with the horseshoe. What he came up with was the world's strongest magnet to date. NYU professors Gale and Draper were familiar with Henry's exploit, and together they made a Henry magnet for the Morse telegraph.

Just back from a failed attempt to obtain telegraph patents in Europe in 1839, Morse finally got around to visiting Professor Henry, who was then at Princeton University.[17] His main concern was to find a way to extend the electrical impulses over long distances. The scientist advised him that "so far as I am acquainted with the minutiae of your plan," the telegraph seemed useful for comparatively short distances. As to extending its range, he would be quite willing to "converse freely" on ways to "modify" the machine. But Henry had no clear answers as to how that might be accomplished.

There was no follow-up to this meeting until 1842, when Vail (whose workshop at Speedwell, NJ, was not very far from Princeton) arranged for Henry to see the telegraph, which had been improved since the initial unveiling in 1837. The scientist was impressed—not least by its unerring ability to send clear signals down miles of wire.[18] Morse contacted Henry for technical advice several times in 1844, during the planning and materials ordering stages of the Washington-Baltimore experimental line. He asked about the relative merits of iron vs. copper wire, among other things. The Princeton professor never failed to send detailed replies.[19]

Henry was also cited in defense of Morse's precedence over Wheatstone in *Silliman's Journal* (Vol. 19, p. 404). The premise of the article was that Henry's 1830–1831 bell ringer experiments predated anything that Wheatstone had done, and those experiments were helpful to Morse, while Wheatstone struggled on without knowing about them. The article was reprinted in the *National Intelligencer*, Vol. 4, p. 32.

Vail's first history of the telegraph (*The Electro-Magnetic Telegraph*) came out in 1845. It featured a detailed account of the science and technology that went into Morse's machine, yet was somewhat dismissive of Henry's contribution to the enterprise. Henry took this as a personal affront and a serious breach of scientific ethics. Assuming that the inventor had vetted the manuscript and therefore approved its contents, he rebuked Morse in letters to colleagues that generated a groundswell of professional opinion that the work should be withdrawn.

Considering Henry's stature in the world of science, and the fact that he had never hinted at monetary compensation, Morse would have been well-advised to cater to his wounded feelings. Vail in fact sent a conciliatory letter on 12 February 1846, which promised that a new edition of the book would set matters straight. It made the additional point that the inventor was traveling in Europe when the offending tome was being readied for publication, and did not have a chance to review it. This was followed by a letter signed by both Morse and Vail, dated 22 July 1846. It was cordial and respectful, but suggested that Henry's satisfaction with a new edition would likely hinge on whatever new information he could submit to Vail, the author. Morse followed up with another reassuring letter on 17 October. There is no indication that Henry sent any "new" information, and when *The New History of the Electro-Magnetic Telegraph* came out in 1847, his anger was redoubled. At some point he reread the Vail letter, and scribbled at the bottom of it: "This letter was written by Professor Morse." With regard to the promised amends, "... this was never done."

All this was just in time for the Morse v. O'Rielly case in Frankfort, KY. The Princeton professor was called by the defense and submitted a deposition. O'R's strategy was to attack the Morse patent, by citing Henry and Dr. Jackson as the true creative forces behind the telegraph.

Only some parts of Henry's statement referred directly to the telegraph. His experiments in 1830-1831 disproved the findings of the British scientist, Barlow, that electrical currents could not travel far along wires. He employed the new Sturgeon-design magnet and a large battery, and found that a strong electrical charge could overcome resistance for some distance along a wire. On the other hand, Henry's deposition expressed the belief that Barlow's "electro-magnetic telegraph" contained the basic features of the Morse telegraph, even though it was only as a research tool rather than a practical invention. He added that his electric bell contained those same telegraphic features. These comments, however, left completely unresolved the distinction between scientific experiments and the practical inventions that such research could give rise to.[20] The Morse side belittled the deposition on those grounds. The defense, however, made much of Morse's contacts with the Princeton scientist, and held these up as proof that Morse was not the sole or even main inventor of the telegraph.

Henry's statement also took up the crucial issue of the origins of the telegraph relay (the combined circuit). The relay was well-named, because it functioned somewhat like runners in a relay race. Henry implied, in very general terms, that the idea of the combined circuit—the magnet and battery connection that revived a dying electrical signal at intervals along the lines— was his. Morse's lawyers asked for proof of his priority on the combined circuit. Where, they asked, had he published a description of a combined circuit before Morse received his patent for one in 1840? The answer was nowhere, because Henry was unable to come up with any verification.

In 1837, Morse had asked him for help on this problem. Henry had, after all, disproved Barlow's claim that wire was all but useless as a conducting medium. He did this by applying a Sturgeon magnet—which was much more powerful than the one Barlow had used.

But now, six years after the Barlow experiment, Henry had nothing new to say. His advice was simply to launch the signal using a powerful magnet. He might have been hiding something—but he had published nothing of importance on the subject. By this time, Morse knew that such a magnet would have to be the size of a house to send a message more than a few miles. Did Henry actually think that he had solved the problem of distance in 1831? He apparently thought so. Dr. Jackson thought so too, and had informed Morse to that effect aboard the *Sully* in 1832, a year after Henry's experiment with the strong magnet.

George Wood, Amos Kendall's assistant at the Washington & New Orleans Telegraph Co. in Washington, D.C., put the matter succinctly in one of his frequent letters to the inventor: "The question is, did Henry invent the Combined Circuit?"[21] The brief prepared by O'R's lawyers claimed that he had (on page 7), despite the testimony of Morse and Gale that it was Morse's idea alone, in 1837, when they were struggling with signals that weakened along miles of wire in the studio at NYU.

O'R lost his case, despite the support of the two scientists, Henry and Jackson. Henry was embarrassed that a panel of judges (in Frankfort) and ultimately the U.S. Supreme Court would discount his testimony. He published an article on what he perceived as deficiencies in the Frankfort decision in *The Telegraph Companion* (Tal P. Shaffner's magazine) in January 1855. The article offered no new insights or information, but only criticized the judges for undervaluing his bell experiment and the Barlow study. A copy was sent to the Franklin Institute in Philadelphia, for inclusion in its scientific archive.

Morse wrote a reply that zeroed in on the issue of the combined circuits. Henry had never mentioned them in print until 1849, after the Frankfort judges rendered their opinion in favor of Morse. Yet his deposition asserted that he had found the "solution" to the distance problem, sometime "before October 1836"—the exact date being beyond recall. He had no particular term or name for his solution, and first used the term "combined circuits" only after Morse's attorney mentioned it in court.

Henry provided a deposition for O'Rielly and the various other parties who challenged Morse's application for a patent extension in 1854—a challenge that went all the way to the Supreme Court. O'R also brought the inventions of Steinheil, Davy, and Wheatstone into his case, claiming that they all predated Morse, whose telegraph was only an amalgam of their telegraphs.

The Supreme Court divided 6-3 in favor of Morse. He was actually elated by the minority opinion, because the three votes against a patent extension only rejected his hardship claim (that patent infringers deprived him of a rightful return on his invention), and not the validity of the patent itself. Both the majority and minority reports of the Court agreed that Morse was the original inventor of the "electro-magnetic recording telegraph." Chief Justice Taney noted, for instance, that "it is evident that the invention of Morse was prior to that of Steinheil, Wheatstone and Davy," and was also markedly different from their various telegraphs. Henry's deposition did not strike the justices as being germane to these main issues.

Henry left Princeton to become the founding director of the Smithsonian Institution, a post that not only reflected but further enhanced his standing in the science community. Morse wrote bitterly that the appointment would reinforce Henry's "pompous pretensions to the Dictatorship of science in this country."[22]

Henry submitted a reprise of the arguments contained in his court depositions to the Smithsonian's board, for publication under the Smithsonian imprint. The board, which consisted of scientific colleagues, approved the paper without any changes. It mentioned new witnesses to the electrified bell experiment, since Morse's attorney had cast doubt on the experiment itself. Henry's improvement to the Sturgeon magnet (the tighter winding of the wire around the iron magnet, and the silk-sheathing wire insulation that kept it from touching the magnet) was cited as an indispensable contribution, without which Morse's telegraph would have been impossible.

Morse wrote a rebuttal and self-published it in pamphlet form. Its main point was to distinguish science from invention. Discovery of the laws of physics was one thing, the invention of a working machine quite another, even if it made use of those discoveries. Admitting that science lay at the foundation of his invention, the inventor noted that his mentor, Professor Dana, was a pioneering researcher in electro-magnetism, too. Dana's work may have influenced Henry, just as Sturgeon and his original magnet influenced both of them. He was especially keen on the point that there was "not one scintilla of evidence" to prove that Henry used a combined circuit in his bell experiment or anywhere else.[23]

There matters rested for a decade. In 1869, the inventor was preparing a report on the recently ended Paris Exposition of Science and Technology, in his capacity as an official U.S. delegate. The account of the telegraphic exhibit included a brief history of his own invention. William P. Blake, a state department official, was assigned to edit the report, and exchanged many letters with the inventor about it over a span of nearly a year. Blake was disturbed about the lack of any reference to Henry, and considered the matter important enough to warrant a personal meeting with the inventor.[24]

In their subsequent talk, Morse pointed out that his report omitted other scientific discoverers besides Henry. It did mention two—Arago and Sturgeon—because he considered their contributions to electro-magnetic research to be of fundamental importance. (They were, in fact, the original pioneers of such research.) The Frenchman Pouillat was omitted, even though his improved magnet anticipated Henry's by several months. (In a later clarification, Morse described the Pouillat magnet as "two horseshoes each wound with 10 thousand feet of covered wire.") Professor Barlow was also omitted, even though Henry's bell experiment was a take-off from the one that Barlow designed in 1824. Soemmering, author of "On the Electric Telegraph" in 1810, was omitted, because his essay was not about an invention, but rather about measuring the strength of currents along a wire. Morse emphasized to Blake that his report was about an invention, not a history of scientific discoveries.

In January 1872, Gale wrote to Morse, who was at his home in Locust Grove, about a new flare-up of the Henry saga. A life-size statue of the inventor was about to be unveiled in Central Park, NY, and F. O. J. Smith had written to the newspapers complaining that Professor Henry, as the true inventor of the telegraph, deserved the honor instead. According to Smith, "Henry stated that electrical currents might be sent through long distances applicable to telegraphic purposes." His letter, which appeared in several of the city's dailies, included a suggested design for a Henry commemoration: a plaque display of his head, surrounded by smaller images of other telegraph pioneers. The smaller images would include Wheatstone, Steinheil, and others, including Morse.[25]

HARRISON GRAY DYAR

During the preparations for the Morse v. Bain proceedings, Dyar (also spelled Dyer) received a request from Bain's attorney for answers to a list of questions, and sent a notarized deposition from his home in Paris, France. In response to one question, describing his prior experience with the telegraph, Dyar swore that it amounted to nothing more than an experiment he had once performed at a race track on Long Island, NY, in 1828.

He gave the following details: A wire was strung all the way around the track, suspended from poles. One end was attached to a battery ("voltaic pile"), and the other hovered over a slowly moving roll of paper (moved by a hand crank). Touching the wire to the battery produced a spark at the other end that left a small burn mark on the paper roll. Dyar's manipulations at the battery end produced patterns of long and short markings. Each pattern matched a letter of the alphabet, in accordance with a code that he had made up.

The statement offered no explanation as to why there was no follow-up. Bain and his attorneys knew, however, that Dyar left the country—apparently for good—shortly thereafter.

Another of the questions Dyar responded to in the "Interrogatories of Harrison Gray Dyer of Paris on the Part of the Respondents" asked how well he knew one Thomas P. Walker.[26] The response was that Walker was his attorney at the time of the telegraph imbroglio, and also an "intimate friend" of long standing. Another question asked if he was aware that Walker was Morse's brother-in-law. The answer was yes.

The "Interrogatories" also asked how familiar Walker was with the telegraph. He knew about it (Dyer wrote), "...in general...but not...with any great precision as to the details..." Another question asked if the respondent knew that Walker was related to Samuel F. B. Morse. Yes, Dyar knew—Walker was the brother of Morse's late wife, Lucretia (who died in 1825). Moreover, both Morse and Walker were residents of New York City when the telegraph experiment took place on Long Island, but neither was present at the experiment.

In response to another question, Dyar remarked, "I had thought it very remarkable that Mr. Morse's plan should be so exactly like my own, especially extending to the mode of reporting the letters of the alphabet, which is identical."

Asked if he was aware of any other litigation relating to the Morse patent, Dyar replied that he had seen O'Rielly's ad in a Paris newspaper, offering a $300 prize for the best essay on the history of the telegraph, but had not submitted an entry.

Asked if he and Morse had ever met, Dyar answered no. He was aware, however, that Morse was in Paris in 1831. His friend Walker had sent him a letter of introduction to Morse, but the opportunity to present it had never arisen.

Dyar further explained that he had been tempted to challenge Morse's patent, but the prospect of a lengthy absence from Paris, plus the expense, put him off. Besides, he was afraid such a move would offend his friend, Walker.

Bain, obviously, was onto something that O'Rielly and all the other patent challengers had never dreamed of. The key connections were, one, Thomas P. Walker; and two, the date. Morse spent months in Paris, copying paintings in the Louvre Museum. On the voyage back to the United States, aboard the *Sully* in 1832, he had little to say about his art work, but talked incessantly with his fellow passengers about the telegraph.

Even before leaving Paris, Morse's roommate, a businessman named R. S. Gabersham, recalled all of a sudden hearing incongruous mutterings about telegraphs. James Fenimore Cooper, the famous novelist and friend of the inventor, who happened to be in Paris at the same time, had a similar recollection. (Both were submitted as sworn depositions at the patent proceedings in Frankfort, KY, in 1848. Coming from friends, of course, their objectivity can be questioned.)

F. O. J. Smith became involved in new telegraph litigation in Boston (Smith v. Joseph W. Clark, 1851), while Morse v. Bain was still pending in NYC. Clark was president of the NY & New England Telegraph Co., a new company just beginning construction, which had obtained the rights to the Bain patent through O'Rielly. The case ended in a standoff, but during the evidence gathering phase, both sides obtained the case file in Morse v. Bain, and found the Dyar deposition in it.

Smith wrote to Morse, asking for Walker's address. He wanted this, he said, because an allegation of a Morse-Dyar connection might crop up in future patent litigation. The inventor replied airily that "drowning men will grasp at straws," and informed Smith that Walker had been dead "for many years."[27]

The Dyar issue was entirely overlooked when Morse's patent extension was challenged in 1854. Dyar's sworn statement, of course, seemed to close the matter. Nobody but Smith found it suspicious, Smith alone being alert to the remarkable coincidence of Morse coming away from Paris in 1832 with the idea of the telegraph in his head, "out of the blue."

No record exists of any effort by Smith to contact Dyar himself, even though his Paris address was in the Morse v. Bain file. In his later attacks on Morse as the inventor of the telegraph, Smith would not mention Dyar at all, preferring to focus exclusively on the scientific contributions of Professor Henry.

The Morse Archive contains no mention of Dyar, either, except for the copy of the Interrogatories. It does contain, however, a printed guest list from a banquet that was arranged by William Fothergill Cooke to honor the inventor. The event took place in London, on 9 October 1856, and the list included the name of one C. K. Dyar.[28]

Nearly two years later, Morse was back in Paris, and a group of American expatriates threw a banquet in his honor at the *Trois Freres Provencals* restaurant. The printed guest list, dated 17 Aug 1858, included a Samuel B. Dyar.[29] Harrison G. Dyar had no relatives by that name.[30] If Dyar lied in his sworn statement, and was accustomed to altering his name on public documents, what exactly did he have to hide?

Dyar, Jr., in the *Preliminary Genealogy*, p. 3, states flatly that his father fled the United States in 1831 to escape prosecution for engaging in a "conspiracy" to convey "secret intelligence by wire." There was a money dispute with an assistant who helped set up the racetrack experiment on Long Island. The assistant filed a civil complaint with a justice of the peace. When this provided no relief, he went to the New York police, claiming that the telegraph project, which involved building a line from New York to Philadelphia, was intended for a criminal purpose.

This brief account parallels comments made, years earlier, by Rev. Charles S. Harrower in his eulogy at Dyar's funeral in Rhinebeck, NY, on 3 February 1875. Harrower referred to "a conspiracy for carrying secret information from city to city." He added that the "Morse language" ought rightly to have been called the "Dyar language," and said the deceased never pursued his rightful claim because of his "noble" temperament.[31]

This is all we know for sure about Dyar's desire for anonymity. No record exists anywhere that sheds more light on it. It is possible, however, to reconstruct a chronology of events that gives us a better understanding of Dyar's self-exile.

In 1827, shortly before the telegraph experiment, the New York Stock Exchange (which was founded in 1817) allowed investors to buy stock with borrowed money (on "margin") for the first time. As a result, the volume of transactions, which previously had averaged barely one hundred per day, increased sharply, and so did the volatility of prices.[32]

Philadelphia was the home of the country's only other (and oldest) stock exchange. Prices of stocks there were never exactly the same as prices for the same shares in Gotham, except perhaps by rare coincidence. If a share of, say, the Yazoo Land Co. was selling for $10 in Philadelphia, and $9.50 in New York, somebody could buy in Gotham and sell in Philly and net 50 cents a share. This, however, was physically impossible. What was needed to make it possible were, (1) speedy communications, before any adverse price changes occurred, and, (2) secrecy—the price spread should be known only to the persons having access to the speedy communications. (Present-day electronic trading relies on (1), but of course (2) is forbidden.)

In 1833, five years after Dyar's experiment, and two years after he fled the country, the New York and Philadelphia stock exchanges were linked for the first time by a semaphore communications system built mainly in the New Jersey countryside. As a result, current prices in Philadelphia could be posted in New York in less than an hour, and vice versa. The desired result was the narrowing of the price spread between the two exchanges, and in this the semaphore succeeded. The system required manned hilltop stations equipped with lanterns and mirrors, and significant tree removal for clear sightlines—a very expensive proposition. This, we propose, was a response to the threat posed by the Dyar telegraph scheme.

Henceforth, Harrison G. Dyar had to steer clear of telegraphs. Yet the telegraph was such a brilliant idea. Why should it go to waste? Enter Thomas P. Walker, and his brother-in-law, the starving artist whose late wife had left behind three small children.

There was an aftermath to this story, which tends to reinforce our hypothesis. Morse had a country estate in the Hudson Valley (near Poughkeepsie) and a house at 5 W. 22nd Street St. in New York City. When Harrison G. Dyar finally returned to the United States, after living in Paris for thirty years, he purchased a house in Rhinebeck, which was about fifteen miles from the Morse estate. He also bought a home in Manhattan on West 24th Street—a short walk from the Morse townhouse on West 23rd Street. In 1865, four years after settling in Rhinebeck, the sixty-year-old Dyar married a forty-two-year-old woman named Eleanora Hannum. Sarah Morse, who also was much younger than her husband, was acquainted with Ms. Hannum. The Dyars visited the Morses at Locust Grove quite regularly. But this greatest of the secrets of Locust Grove remained a secret.

DR. CHARLES T. JACKSON

Morse stated that the idea of a telegraph suddenly entered his head "…in a conversation with one of the passengers" aboard the *Sully*, en route from Le Havre to New York, in 1832. They were talking about Benjamin Franklin's kite experiments with lightning, a topic that every American school child knew about. Morse had learned more about the subject of electrical conductivity from Professor Benjamin Silliman, when he attended Yale between 1805 and 1810. Silliman actually did classroom demonstrations with electrified wires, energized by an Ampere—Volta type battery that he had made himself. The students would join hands, one would touch the wire, and all would feel a jolt. Colleagues quipped that somebody had finally found a way to keep students awake in class.

Silliman's study of the famous "Weston Meteor" was also brought up aboard the *Sully*. This meteor had landed a few miles from Yale during Morse's sophomore year, in 1807. The students heard the impact during a chapel service. Silliman brought his students (including Morse) on field trips to the meteor. He always brought along a piece of magnetized iron—probably a by-product of Franklin-type experiments with lightning. Wherever the natural magnet stuck to the surface of the meteor, the professor explained, iron must be present. Everyone was amazed.[33]

The passenger Morse referred to was Dr. Jackson, a physician and geologist. Jackson was familiar with chemistry, and had used natural magnets to detect iron deposits. More discussions about electricity, electro-magnetism—and the feasibility of a telegraph—followed. But according to the inventor, they all took place after he had broached the subject of applying science to the creation of a practical recording telegraph.

At one point, he asked Jackson's opinion about the best way to use electricity to make an impression on paper. Jackson thought that paper treated with some sort of chemical and salt mixture might have possibilities, but experiments would have to be done to find out what worked best. Also mentioned, as a possible alternative to this, was an electro-magnetic marker—a pen or pencil guided by an electrified arm. At the end of the voyage, the two men agreed to get together again on this intriguing telegraph idea, but they never did.[34]

Jackson read about Morse's inaugural demonstration of the telegraph at NYU in 1837, in a Boston newspaper. He immediately wrote a letter to the inventor, noting an "oversight" in the story—its failure to mention Jackson as the "co-inventor." A reply to Jackson acknowledged his kind advice aboard the *Sully*. It went on to acknowledge the influence of Professor Silliman, with his demonstrations of electricity and magnetism at Yale; and also the inventor's brief but close association with John Freeman Dana, a Columbia University science professor who did research in electro-magnetism. The letter also mentioned his NYU colleague, Professor Gale, acknowledging him to be a "mutual inventor, on the score of aid by hints." Jackson surely deserved credit for his advice, but these others—Silliman, Dana and Gale—deserved even more.[35]

When Jackson expressed his strong dissatisfaction with this response, Morse solicited letters from the *Sully*'s Captain Pell and fellow-passengers. Pell's reply was quite explicit. He recalled Morse talking constantly about his telegraph idea, and recalled seeing a picture of the invention in a newspaper, five years later, that distinctly reminded him of sketches Morse made aboard ship.

William Rives, the American envoy in Paris who was also returning home aboard the *Sully*, made similar comments, as did the several other passengers who responded to the inventor's request for statements. None of the responders backed up Jackson's story, and he did not solicit any affidavits from the passengers who had not answered Morse.

The doctor's response to all this was a letter in the *Boston Post* claiming not to be the co-inventor, but the sole inventor of the telegraph. This letter offered as justification the premise of (as he put it) "first idea...entire invention." The letter was reprinted in the *New York Courier & Inquirer* (12 January, 1839), while the inventor was in Europe seeking patents in England and France. Jackson also wrote to the *New-York Observer*, the church news weekly that was published by Morse's brothers, Sidney and Richard. They sent it on to the inventor, who was then in Paris, without publishing it. The tone of this latest missive was more urgent than before, with much underlining (double and triple underlining, in places). It concluded: "I furnished him with all the necessary data for making the electro-magnetic telegraph."[36] Morse saved the letter, with a comment attached: "...looseness of accuracy...makes his statement false." He then wrote another rejoinder to Jackson.

Two months later, the *Post* letter found its way into a Parisian science journal, the *Comte Rendu*. It had been forwarded by a Belgian clergyman, Abbé Magnot. He was acquainted with Charles Wheatstone, inventor of the

British telegraph, and Morse's French patent application was pending. (Magnot was also aware of Morse's very public efforts to stem Catholic Irish immigration, back in New York.)[37]

The *Comte Rendu* publication was followed shortly by a letter Jackson sent to Elie de Beaumont, a member of the French Academy. These two had met at a conference in Paris devoted to recent developments in electromagnetic research—the conference Jackson was returning from when he met Morse aboard the *Sully*. The letter included the following comment: "I regret to see that Professor Morse has appropriated my electro-magnetic telegraph. I explained the instrument to him in full…The invention which he has shown to them belongs to me wholly."[38]

Beaumont had recently witnessed a demonstration of Morse's telegraph at the French Academy and, like all the other members, was highly impressed. He gave a copy of the letter to Morse. There was no discernible reaction against the inventor from all this. Indeed, his patent was soon granted.

Little more was heard from Jackson until 1844, when Ezra Cornell, fresh from the construction of the Washington-Baltimore experimental line, set up a telegraph exhibit in Boston at the behest of the city council. Dr. Jackson wrote to all the local newspapers, stating that public attendance at this exhibit would lend dignity to a base crime. He urged a boycott.[39]

After another hiatus, during which Jackson secured an appointment as state geologist of Michigan, his claim resurfaced in the *Detroit Daily Advertiser*, on 18 May 1847. On receiving a clipping of this letter, Morse forwarded to the *Advertiser* his by now standard response: Jackson never actually made and tested a telegraph, never described the mechanics of a telegraph to Morse, and never took issue with the sworn statements of the *Sully* Captain and passengers for the obvious reason that they refuted his claim.

A year later, the inventor's brother, Sidney, happened upon an article in the *Revue Scientifique* while on a visit to Paris. It was devoted to the history of the telegraph, and stated that the Morse telegraph operated on the "même principes" as the Steinheil and Wheatstone machines. More alarmingly, it mentioned the Jackson claim, but without comment.

Despite its glaring errors on matters pertaining to Morse, the *Revue Scientifique* article was very informative on the subject of telegraphic forerunners. It described the endeavors of LeSage in 1774, Lomond and Beaucourt in 1787, Reiser (a German) in 1794, Cavatto and Salva (1795-1796, both Italians), and Ronalds (an Englishman) as comprising the "premier epoch" of the telegraph. The "deuzieme epoch" saw the exploits of Soemering, Coxe, and Schweiger in 1810–1811, Scholling, Gauss and Weber, Steinheil, Amyet, Mason and Brequet, Davy, Cooke, Wheatstone….and Morse. One may wonder how an awareness of this small army of competitors might have affected his resolve, during the years 1832–1837, when he painstakingly put together a working machine in the confident belief that he, and it, were *sui generis*. In the article, the only thing that distinguished him from the pack was not the superiority of his machine, but the fact that he was the only one whose originality was being contested.

In the showdown between Morse and O'Rielly over the latter's unauthorized use of Morse machines, Morse et al. v. O'Rielly et al., in Frankfort, KY, (1848–1849), O'R obtained a sworn statement from Jackson in his counterclaim that the Morse patent was invalid. The statement repeated Jackson's claims, but failed to address Morse's familiar rebuttals to those claims. This was brought to the attention of the judges, and the rebuttals were then hammered home again, one by one. The net effect was injurious to O'Rielly's case. The final judgment against O'R explicitly rejected all of Jackson's assertions.

In 1850, Dr. Jackson was fired from his job as state geologist of Michigan. Morse seized the opportunity to write a pamphlet about his adversary's character, under the compendious title, "Full Exposure of the Conduct of Dr. Charles T. Jackson leading to his discharge from government service and justice to Messrs. Foster and Whitney, U.S. Geologists."

It described how the two gentlemen, Foster and Whitney, were driven to resign their field survey positions in Jackson's department. They blamed him for excessive absenteeism. Backlogs of geological data piled up in his office, and then were often misfiled or lost. In addition, they claimed that he hired untrained "pupils" to do field work, which generally had to be done over. These students paid tutorial fees to Jackson out of the government wages for "geological field training." A state investigation determined that Jackson was guilty of "neglect of duty." He agreed to resign, and when he withdrew the resignation, was summarily fired.

Morse wrote another pamphlet in 1852, titled "Full Exposure of Dr. Charles T. Jackson's Pretensions to the Invention of the American Electro-Magnetic Telegraph." In 1853, Salmon P. Chase, who had been O'Rielly's attorney at Frankfort, wrote a short book about the case. In his closing argument, the only item from Jackson's deposition that he used was the assertion that the inventor had no scientific background, and therefore could not have invented the telegraph on his own. He did not address Morse's rebuttals against Jackson at all, including the inventor's response to that particular claim, even though he had a copy of Morse's first anti-Jackson pamphlet in his court file.

A point that Morse never did address, in his pamphlet or in court, was Chase's observation that Jackson was transporting an electro-magnet and battery on the trip on aboard the *Sully*. Why he had these items was not explained. The defense wanted to leave the impression (and never claimed explicitly) that they were for an electro-magnetic telegraph. In fact, the doctor's interest in electro-magnets was related to his geological work, primarily in the area of detecting iron deposits. He had attended lectures at the Sorbonne in Paris on the recent electro-magnet experiments for that purpose.

In his private notes, the inventor wrote, "He explained the mode of obtaining a spark from the magnet. I suggested the telegraph...." Whether or not Jackson actually performed a demonstration aboard ship is unknown. It is not mentioned in any of the affidavits Morse obtained from people aboard the *Sully*.[40]

In his pamphlets, as in his testimony at Frankfort, Morse challenged the claim that he lacked any scientific or technical background and in particular, no acquaintance with electro-magnetism, as Jackson asserted.

An affidavit from Professor Silliman was entered into evidence. It detailed the demonstrations of electric conductivity, in classes that Morse attended, with a "Cruickshank battery and voltaic pile." At the time, in 1809 and 1810, such experiments were available nowhere else in the United States but in his classroom at Yale. Silliman, moreover, became a family friend after the inventor's parents moved to New Haven in 1824. Morse painted his portrait, and gave it to him as a present. The professor was only twelve years older than his former student, and the two of them often went on long walks together when the inventor was in town.

By then, *Silliman's Journal* was the premier American science periodical. It was, among other things, the best source of information on recent research in electro-magnetism. Professor Sturgeon's breakthrough—his large horseshoe-shaped cylindrical magnet surrounded by a coil of copper wire—was first reported in the United States by *Silliman's Journal*, (Vol. 19, 1831, p. 329).

The very next issue of *Silliman's Journal* contained an article on a British scientist, Sir Humphrey Davy, who was also doing experiments in electro-magnetism. The story included a reference to Professor Joseph Henry's new magnet and his bell-ringing test in Albany NY. It also noted that Silliman himself had just installed at Yale "the most powerful magnets ever," which (at eighty-two pounds each) could "sustain over a ton." The article went on to state that on 12 Feb 1830, a Mr. Ritchie exhibited to the Royal Academy in London, "the electro-magnetic telegraph proposed by Ampere." Copies of both editions of the *Journal* can be found in Morse's Letter File for that year, although it is almost certain that he put them there at a later date.[41]

Silliman did not become a role model for Morse, however. The fact that Morse became an artist rather than a scientist attests to that. Yet it is undeniable that the Yale professor provided the spark that set the inventor off, later on, on his quest for an electro-magnetic telegraph. From the very beginning of his travels, after graduating from Yale, Morse expressed in his letters home an overbearing loneliness that could be measured in time and distance. In a letter to his parents, for instance, he once wrote: "I wish that in an instant I could communicate...but the three thousand miles are not passed over in an instant."[42] When his wife died suddenly while he was away from home, in 1825, the news didn't reach him until the day before her funeral. Things he learned in Silliman's classes, years before, kindled his hopes that the obstacles of time and space could be overcome by means of science and invention.

Morse had an inventive streak, even before he thought of the telegraph. In 1817, with his brother Sidney, he made a water pump, "which we hope will be profitable."[43] This occupied much of his spare time for several years. A large model that was intended for fire pumpers was built at some expense, along with a small model for domestic use. Each was made at a Boston workshop run by a man named Dearborn.[44] The big engine was priced at $200, the small one at $20. The two brothers sought investors, applied for a patent, and even hired a Mr. Hart as their business agent. They placed newspaper ads in 1818, which generated some orders, accompanied by down payments.[45] These deposits did not cover the cost of manufacture, however, and the two Morse brothers soon faced a "cash flow" problem. People began demanding refunds when pumps were not sent promptly. Morse lamented, "surely an inventor earns his money hard."[46]

In 1823, Morse filed a caveat with the Patent Office for a marble-cutting machine. He had built a "small working model" and promised to file detailed specifications in a few months. Precisely where his interest in marble came from (and pumps, for that matter) is a mystery. These endeavors, though, reveal that long before he thought of the telegraph, Morse tried to create new mechanical devices, and thought of himself as an inventor.

In 1827, Morse attended a series of four lectures on electro-magnetism, presented by Professor John Freeman Dana at Columbia University. That Morse travelled far uptown to all four lectures is revealing in itself. More important was the relationship that he developed with Dana. The inventor was obviously fascinated by electro-magnetism, and Dana's expertise in this area was unsurpassed. Among Americans, only Professor Joseph Henry was his equal or master in the field. In 1830, however, Dana's life was cut short by a stroke. He left behind a set of notebooks detailing all of his research. These would have been of considerable use to his scientific colleagues, but somehow his widow entrusted them to Morse. A quarter century later, she asked for their return, because her daughter wanted them. Mrs. Dana was disappointed when the inventor sent only copies, written in his own hand, and claimed that the original notebooks were no longer in his possession.[47]

In 1865, Morse learned that Dr. Jackson had been involved in another invention (or discovery) issue. A former medical student of his, Dr. William Morton, pioneered the use of ether as an anesthetic, in surgeries performed at Boston hospitals in 1846. Jackson claimed that the idea was his, and Morton filed a protest about Jackson's claim with the local medical society. An investigation ensued, and Morton was declared to be "the practical discoverer" of ether anesthesia.[48]

Morse contacted a Cambridge MA attorney whose name had appeared in an article about the case. The attorney, R. M. Dana (no apparent relation to the late scientist), replied that he had compiled a "thick volume" on "Morton's struggles." He also happened to recall "Fog" Smith's patent infringement suit against the House telegraph, years earlier in Boston. House had brought Jackson in as a defense witness, but the judge was totally unimpressed with his testimony. Dana further revealed that the famous Harvard scientist, Louis Agassiz, had advised the panel of investigators on the protocols for attribution in scientific discoveries. Agassiz, he said, characterized Jackson's ether claim as irresponsible, even irrational.[49]

The inventor, buoyed by these revelations, decided to update his anti-Jackson pamphlet. He completed this while travelling in Europe the next year (1868), and had it printed in Paris. Copies were sent to the *Comte Rendu* and the *Revue Scientifique*, which had aired Jackson's claims nearly two decades earlier.

In 1871, the last full year of his life, Morse was working on yet another rewrite of this pamphlet.

IN CONCLUSION

None of these five stories reveals a viable alternative inventor of the telegraph, although Gale and Vail—Vail especially—may qualify as co-inventors. This raises the question of where to draw the line between a helper or collaborator and a fully-fledged "co-inventor." Perhaps we need a new vocabulary word to describe their role. At any rate, Morse acknowledged both with small, fractional shares of the patent. Did Vail deserve more than he got—say a quarter share? His archive at the Smithsonian Institution is entirely opaque on the matter. It displays no privately-expressed resentment against the man he steadfastly acknowledged to be the driving force behind the entire project.

Henry's story deals with the distinction between science and technology. Morse's view of the matter, in his response to claims that the eminent scientist Joseph Henry invented the telegraph, seems entirely valid. Henry's experimental designs were highly suggestive of practical uses, but they were not practical inventions in themselves, and Henry had no pretensions to being an inventor. The dispute between the two men, which was never resolved, arose from Vail's *History of the Telegraph*. Vail slighted Henry's absolutely indispensable scientific contribution. Henry's response was intemperate, but Morse simply dug in his heels, in a reactive response that was also protective of his protégé, when he should have been diplomatic. He was, after all, dealing with a renowned scientist. This was an odd and sad development, considering that the inventor's long-term and highly successful leadership of the National Academy of Design is a testimonial to his interpersonal skills. He was very good at attracting patronage for the institution, and established many close and enduring friendships in the process.

How important is the person who provided the inventor with the idea for his invention? Harrison Gray Dyar did that, and supplied the Morse code as well, which was vital to the success of the Morse Telegraph. No Dyar, no Morse Telegraph.

Dr. Jackson, on the other hand, believed that Morse invented his telegraph only because he convinced Morse that such an invention was now within reach, thanks to recent advances in the study of electro-magnetism. (Recall that Jackson was returning from a conference on electro-magnetism that was held in Paris.) This, he contended, qualified him as the inventor, or at least (as he initially proposed) co-inventor. Morse could have deflated Jackson by revealing his relationship with Dyar. Never mentioning Dyar in his intense and ongoing effort to refute Jackson, once and for all, raises a valid question. Did Morse actually meet Dyar during his Paris sojourn in 1832? The evidence that he did is very strong, even if it is circumstantial. But that leads to the further question: why did Dyar remain in the shadows? After deciding that it was safe to return to the United States, he could have put forth a claim that was stronger than Jackson's, based on his invention of the binary (Morse) code and the fact that he actually built an experimental telegraph before anybody else did. The fact that by then the "patent wars" had already been fought all the way to the Supreme Court was certainly not encouraging for such a belated claim. But the ultimate answer lies in the bond of secrecy that existed between him and Morse. Dyar did not wish to be exposed to any renewed legal harassment from the people who had hounded him into exile. Moreover, had he returned prior to the definitive Supreme Court decision that was written by Chief Justice Taney, he might well have been hailed to the witness stand and subjected to intensive questioning by clever lawyers.

It must be said on Dr. Jackson's behalf, though, that he certainly provided Morse with strong and authoritative encouragement during their fateful confabulations aboard the *Sully*. He was wrong when he assured the inventor that an electric current could travel down a wire ad infinitum, but that was a happy error for Morse, who might have been turned aside by the challenge that lay in store on that score. Had Jackson shown some forbearance, he would surely have been entitled to some public recognition by the inventor, and a material reward as well. Jackson—and Henry—should have been Morse's friends, not his enemies.

NOTES

1. Articles of Agreement, 1845, by which Alfred Vail and Morse's business associate, Francis O. J. Smith, were also awarded patent shares.

2. Cf. O'Rielly, et al., v. Morse, U.S. Supreme Court deposition dated 15 April, 1850.
3. M—Gale, 14 April, 1860.
4. Gale—M, 8 October, 13 November 1857.
5. Letter dated 1 March, 1843, Vail Telegraph Collection, Smithsonian Institution, Letters (Correspondence).
6. Ref., for instance, Vail—M, 16 October 1857.
7. Vail—M, 21 September, 1848.
8. M—Vail, 15 June, 1854.
9. Initially they were made at the Speedwell works, and nowhere else.
10. George Vail—M, 25 July, 1849.
11. M—Stephen Vail, 24 May, 1861; 4 November, 1862.
12. Mrs. Vail—M, 14 March, 1864.
13. M—William Vail, 13 June, 1864.
14. Cummings Vail—M, 13 June, 1871.
15. M—Z. Allen, February, 1856 (no day given); Cf. also M—S. C. Walker, 3 January, 1848.
16. Cf. Patent Office—M, 15 March, 1853.
17. M—Henry, 24 April, 1839.
18. M—Smith, 16 July, 1842 describes the visit. Cf. also Henry—M, 24 Feb., 1842, in which the scientist wrote, "…I most sincerely hope that you will succeed…"
19. Cf. Henry—M, 17 April, 1843. This contrasts sharply with the tone of the letters cited in the next paragraph, written after Henry read Vail's book on the invention of the telegraph. The Henry and Vail correspondence folders are located at the Smithsonian Institution.
20. Kendall—M, 1 February, 1853.
21. Wood—M, 4 November, 1853.
22. M—Kendall, 5 December, 1856.
23. Cf. Kendall—M, 6 August, 1857; M—Sidney Morse (brother), 31 March, 1859.
24. Blake—M, 19 August, 1869.
25. Gale—M, 12 January, 1872.
26. Insert in the Morse Letter File, 1850 Folder. The interrogatories were subpoenaed by the defendant in a patent suit filed by the inventor's erstwhile partner, "Fog" Smith, against a rival telegraph operator. (Francis O. J. Smith v. Joseph W. Clark, Federal District Court, Boston, 1850.)
27. M—Smith, 12 August, 1851.
28. Cooke and Wheatstone had once worked together on a telegraph. The guest list appeared in the *New-York Observer*, the religious weekly owned by the inventor's younger brothers, Richard and Sidney.
29. M Clipping File, September, 1858.
30. Harrison Gray Dyar, Jr. *Preliminary Geneaology of the Dyar Family*. Washingfton, D.C.: Gibson Brothers, Printers, 1903.
31. Charles S. Harrower, "In Memoriam. An Address by Charles S. Harrower at the Obsequies of Harrison Gray Dyar, in Rhinebeck, New York, 3 February 1875. New York" E. H. Jones & Co., Printers, 1875 (Library of Congress Electronic File).
32. Jerry Markham, *Financial History of the United States*, Vol. I. Armonk, NY: M.E. Sharpe & Co., 2011. pp. 123, 157, 160, 163.
33. Michael B. Chandos, *Benjamin Silliman: a Life in the Young Republic*. Princeton University Press, 1989. p. 221–224.
34. Morse, Personal Notes, 1855. The inventor first tried a chemical approach, and then opted for the marker.
35. Jackson—M, 10 September, 1837; M—Jackson, 18 September, 1837.
36. Jackson—*New-York Observer*, 19 January, 1839.
37. *Comte Rendu*, 4 March, 1839.
38. Morse, General Correspondence Folder, May, 1839.
39. Cornell—M, 23 November, 1844.
40. Cf. file in Morse archive titled, "Notes on the Telegraph and Pending Patent Controversy."

41. *Sillliman's Journal*, Vol. 20, July 1831, p. 143.
42. M—Parents, 3 October, 1811.
43. M—Lucretia, 29 March, 1818.
44. Ref. M—Parents, 22 December, 1818.
45. Cf. Jos. Tompkins—M, 12 March, 1818.
46. M—Lucretia, 17 November, 1818.
47. M—Matilda Dana, 7 February, 1855.
48. M—Richard Morse (brother), 8 July, 1865; M—J. S. Whitney, 14 July, 1865; M—Dr. Henry Bowditch, 14 July, 1865.
49. R. M. Dana—M, 13 January, 1867.

Chapter Five

The Great Man Revered and Reviled

Morse had a strong interest in politics and public affairs, and even ran for political office a couple of times—once before inventing the telegraph, and once after. Neither attempt was successful. The main reason for this was that he associated himself with highly controversial positions on the issues of the day. Morse was deeply committed to these causes, and traded on the fame and fortune that greeted him in mid-life to advance them. Inevitably, his name became associated with controversy. A curious juxtaposition of fame and notoriety was the result. In an era that had no shortage of larger than life heroes and villains, it was something to behold, and it sheds an interesting light on the raging issues that threatened the very survival of the nation.

Morse spent the entire duration of the War of 1812 living in London as an art student. In 1813, while the British were fighting the Americans for control of the Great Lakes, the young artist, an enemy alien, won a major prize at the annual art exhibit of the Royal Academy. His Gold Medal for sculpture was presented by the Duke of Cumberland, a member of the royal family, and the artist was applauded by a distinguished assemblage of his country's enemies.

Morse had been cautioned by his parents to remain on his best behavior when the war broke out. There was no thought of trying to leave the country. Mail contact was slow but steady, thanks to third party arrangements—travelers and captains aboard neutral ships who were ever-ready to carry strangers' mail. The young artist was obviously unconcerned that the British authorities might read the letters he wrote home. These were invariably filled with bellicose anti-British rhetoric, and contrasted sharply with his parents' occasional and somewhat ambivalent comments on the war.

His father, Jedidiah, admonished in one letter that, "The war is very unpopular in this country."[1] Jed was correct only about his native New

England, which fretted about the loss of its lucrative British trade. More than once, in letters to her son, the artist's mother complained that he rarely wrote about his personal affairs, but instead penned page after page about international politics.

The letters home sometimes contained interesting observations about British perspectives on the war. For instance, the young artist reported heavy coverage in the local newspapers of anti-war protests in New England, and especially of the Hartford Convention, which in 1813 nearly voted to take New England out of the Union and make a separate peace. The artist also said the London newspapers described the United States as "a nation of cheats, sprung from convicts..." Britons overall, he thought, had a "haughty...overbearing" attitude toward Americans. As the war dragged on, however, Morse reported that the British gradually stopped looking down on the Americans, and developed a more hardened and wary attitude.[2]

The Napoleonic Wars were in progress during that time, also, and the young art student was decidedly opinionated about Napoleon Bonaparte, as were most (if not all) other people living in Europe at the time. He thought the Emperor's surrender in 1814 a "most glorious event," and was part of the crowd in Piccadilly that cheered a parade of dignitaries, including King Louis XVIII of France, as it rolled by in ornate coaches, en route to a victory reception at the royal palace.[3]

Jedidiah Morse was anti-war, but he had an even stronger interest in the new movement to abolish the slave trade. He corresponded with William Wilberforce, the English clergyman who was the virtual founder of the anti-slave-trade movement in England. *The Panoplist*, a monthly religious journal that Jedidiah edited, printed some of the letters he received from the English reformer. Since New England shippers had a long record of participation in the trans-Atlantic slave trade, Wilberforce was thankful that his colleague was carrying the message into that region.

On reading that Wilberforce had been invited to address the House of Commons on the subject, young Morse decided to attend. Sitting in the visitors' gallery, he heard the crusading churchman call for the "universal abolition" of slavery itself, once the slave trade was over. Shortly thereafter, thanks to a letter of introduction from his father, he was invited to a breakfast with the reformer. The two never met, however, as Wilberforce cancelled on a day's notice, and did so again after rescheduling. Had he known that the young man would later become an activist "Copperhead" Democrat—a Northern pro-slavery advocate—he might have made a more robust effort to meet him.[4]

Late in 1823, Morse was in Washington painting the House of Representatives. The artist had arrived just in time for the New Year festivities, and described in a letter home the "drums, horns, kettles, fifes, from sunset on...:" that livened up the streets of the capital. It was a familiar scene. He had been there exactly two years earlier, for a portrait sitting with President James Monroe. It had been commissioned by the City of Charleston, SC, on the occasion of a visit by the new president, who only had time for one painting session during his brief stay. Monroe was still in office, so the artist left his calling card at the White House, and in return received an invitation to the president's New Year's Day Reception.

A week later, Morse was offered a post as clerk in the just-forming U.S. diplomatic legation to Mexico, which had become an independent country in 1821. This sudden opportunity immediately set him to dreaming about exploring Mexico's art treasures, and becoming a dealer in Mexican art.

The first sign of trouble in this new venture came when he was told to bring his own bedding. This occasioned an urgent letter home, for a mattress and pillow to be dispatched by packet boat. Word then came down that the mission's departure to Mexico was being delayed, because of some civil disturbance that was occurring in the new nation. It also transpired that, while the government was responsible for the trip to Mexico City, there was no provision for salaries. (Note: Diplomats in those times were assumed to be motivated by a desire to serve their country, gratis, while pursuing personal interests abroad, and the state department had no budget for diplomatic salaries.)

Morse began looking for another federal sinecure, but nothing came up. Overtaken by news that Mexico was a land of "anarchy and assassinations," the artist then headed for home. A year later, after wandering around New England trying to sell portraits, he settled in New York.[5]

In 1826, when the Marquis de Lafayette visited the city, Morse was commissioned to paint his portrait. The aging Frenchman was impressed by the artist's political views, and they struck up a correspondence. Five years later, when Morse was in Europe on the sabbatical that the National Academy's trustees paid for, he got to meet Lafayette again, at his home outside Paris.

The aging soldier-statesman was currently occupied with the fate of refugees from the Polish Revolt of 1830. Many were living at loose ends in Paris, and Morse was asked if he could help find a home for some of them in America. The artist promised he would do everything possible. Once home, he enlisted his friend, James Fenimore Cooper, in the cause. The two were accustomed to talking politics at coffee house gatherings near the academy in downtown Manhattan. Cooper's best-selling frontier novels were also very well known around Europe, and there was a hope that his name, attached to a humanitarian cause, might carry some weight.

Subsequently, even though the refugee problem gradually subsided, the correspondence between Morse and Lafayette maintained a brisk regularity. The letters, which were always devoted to the struggle for democracy, became longer. Morse often included newspaper clippings relating to the developing controversy between North and South. These interested Lafayette considerably. He was well aware that General Washington had been a slave owner, and he had seen slaves at Yorktown, including some, freed by the British, who were being processed for return to their owners after Cornwallis's surrender. Lafayette's view of the sectional controversy was that the spirit of compromise must prevail on both sides. With the permission of the aging Frenchman, Morse sent copies of one letter explicating this position to newspapers around the country. Lafayette thanked him for doing this.[6]

A letter in the Morse Archive dated 20 June 1834 has the following note attached to it: "Last letter from Lafayette, on his deathbed, too weak to sign."

The same topics occupied much of the artist's correspondence with his friend, James Fenimore Cooper. One letter summed up his views concisely: "America is the stronghold of the popular principle, Europe the Despotic.... (including) the heartless, false, selfish system of Great Britain, the perfect antipodes of our own." A subsequent letter addressed the growing sectional crisis: "Matters are serious at the South...they are mad there...hotheaded."[7]

About the time he wrote these words, the first boats crowded with Irish Catholic immigrants began arriving at the Castle Garden processing center on the southern tip of Manhattan. The artist was startled the first time he came within earshot of a thick Irish brogue on the streets of New York. It recalled to him a long-ago foggy morning when Irish fishermen came alongside the "Lydia," off the coast of County Cork, to peddle freshly-caught herring and cod. The art student, who was en route to a three-year stint at the British Academy, laughed then at their hearty brogue.[8]

Morse wrote letters to the newspapers questioning the wisdom of admitting these people. In 1835, he composed an anti-immigrant pamphlet and scraped up enough money to have several hundred copies printed. The new Native American Party, which had a significant following in Gotham, recruited Morse as a member, and nominated him to run for mayor. They were banking on his standing as head of the city's premier cultural institution, the National Academy of Design, whose board of trustees included some prominent citizens.

The artist-turned-politician finished last in a field of four, his effort not helped by the fact that the day before the election, the papers reported that he had dropped out. None of the city's elite provided him with any backing. Prepared to make a second run in 1840, Morse backed off and then quit the Native American Party when it began mulling support for the Whig Party candidate, instead of running a candidate of its own.

Morse's 1835 pamphlet was titled, "Imminent Danger to the Free Institutions of the United States through Foreign Immigration and the Present State of the Naturalization Laws." (Pseudonymous authorship: "An American," NY: E.B. Clayton, publisher, 1835.) It was paid for out of his own pocket and offered free at cooperative newsstands, even though his sporadic earnings from student tuition at NYU barely allowed him the price of one square meal a day. The response to a request for donations on the back page, however, produced a steady if modest flow of cash that paid for successive reprints.

The message was religious rather than ethnic, and it made two points: (1) that Catholics were monarchists because they owed unquestioning obedience to the Pope, and, (2) that the wave of Irish Catholic immigration, if continued unchecked, would turn the United States into a province of the Papacy.

Shortly after "Imminent Danger" made its first appearance, a German immigrant named Henniger Clausing contacted the artist, asking for protection against persecution by the Jesuits. Morse tried to interest the newspapers in his case, without success, and then induced his new friend to begin writing his own story. At this point, a newspaper reported that Clausing was the subject of a manhunt in Germany. He had been charged with killing a priest during a student riot in Heidelberg, Germany, in 1830, and barely escaped arrest while hiding in Paris.

Clausing admitted to being a fugitive when Morse confronted him with this story, but insisted that his pistol had gone off by accident. A few days later, Clausing's name was on the front pages. He had shot himself dead in a Bowery hotel room.[9]

This episode did not dampen Morse's anti-Catholic ardor. Shortly thereafter somebody sent him a hand-written manuscript titled, "Confessions of a Catholic Priest," with a request to write an introduction for a new English edition. (It had originally appeared in France, in 1832.) The story detailed the illicit sexual and financial peculations of an anonymous village priest. The introduction was promptly written, despite Morse's total ignorance of its provenance. "Confessions" duly appeared in a private printing by the anonymous sponsor. Nobody ever found out who the priest was, or even who sponsored the publication, but the introduction bore the signature of Samuel F. B. Morse.

In 1840 Morse was approached by the New York consul of the Kingdom of Sardinia. (Sardinia was commonly known as Piedmont, and its capital was Turin, on the Italian mainland.) The diplomat expressed interest in the inventor's views on Catholicism and, noting that the artist had sojourned in Rome from 1829 to 1831, wanted to know if he had any particular notions about the Church in Italy.

Morse regaled him with vivid memories of the Papal succession that followed the demise of Pope Pius VIII (who died on 30 November 1830). The artist was sketching in a Vatican gallery when the news broke. A crowd quickly gathered in St. Peter's Square, and grew exponentially with each passing hour.

After days of anticipation, the hoped-for whiff of smoke wafted out of a chimney above the room where the College of Cardinals conducted its deliberations, and the new Pontiff, Gregory XVI, duly appeared on a balcony to bless the cheering throngs in the square below. Morse was on hand for much of this drama, and he also had a curbside view of Gregory's ceremonial tour of the city the next day. He recalled being surrounded by "a crowd of the lower orders," and as the Papal coach trundled by, cries of "Papa" rang out, and people dropped to their knees. At that point Morse felt a rude hand slap the back of his head, which sent his hat flying. He turned and saw a policeman glowering at him, and pointing at his head. Morse had failed to remove his hat.

The consul wanted to know if Morse would be interested in being a sponsor of a new Protestant missionary project in Italy. This was an odd request coming from an Italian diplomat, and he was quick to explain that it was not an officially sponsored enterprise, but only the work of non-Catholic Italians like himself. (Note: Twenty years later, in 1860, Piedmont invaded and annexed the Papal States and in 1870, completed the unification of Italy by seizing the city of Rome. For their efforts, the King and all of his senior ministers were excommunicated.) Morse said yes, and soon was named a trustee of the newly-incorporated American Foundation for the Promotion of Religious and Scientific Knowledge among the Italians.

The two men drew up a charter together. It included an organizational chart outlining a semi-autonomous headquarters in the capital of each independent state (Italy not being a united country at this time). Below these were "decades"—urban administrative centers—and "centenaries" in the smaller towns. Presiding over this elaborate hierarchy was a central Directorate located in Turin.

The inventor's notes on this charter include a draft for a speech, to be delivered at an inaugural meeting. Intended for an audience of American supporters, it began, "Popery has reared its head boldly in our midst…and strides over the land…" It cited Italy, "in its present oppressed and degraded state," as the image of America's future if nothing was done.[10]

The Philosophical Italian Society was founded shortly thereafter, and became the public face of the Foundation. Morse delivered his speech at its founding session in NY, and was elected president of the Society. Three years later, the Foundation and the Society were subsumed into the American and Foreign Christian Union.

Morse was literally penniless when he first became involved in this endeavor. But after 1849, with income from telegraph royalties flowing in, he became a financial mainstay and active fund raiser for the Italian missions. His association with this group would continue for the remainder of his long life.

The most notorious of Morse's anti-Catholic adventures was also his last. It began in 1855 when Bishop Michael Spaulding of Louisville, KY, wrote a letter to the editor of the *Cincinnati Enquirer*, regarding Morse's "Imminent Danger" pamphlet. Morse had revised and reissued it, and somehow at least one copy reached Cincinnati and the *Enquirer* ran a story about it. The article was basically a summary of the contents, and included the now-famous inventor's citation of Lafayette, as having warned that the Catholic Church was a threat to American democracy.

The bishop's letter expressed a strong doubt that Lafayette had ever said such a thing, and demanded proof from the author or a retraction.

Morse was notified by the *Enquirer*, and offered a newspaper clipping from the *Somerset* (NJ) *Whig* as proof. It quoted from an alleged private letter by Lafayette to an unidentified person, to the effect that: "If ever the liberties of this country are destroyed, it will be through the influence of Roman Catholic priests."[11]

The inventor's reply was duly printed, and elicited a quick response. Spaulding called the Lafayette letter "apocryphal," and pointed out that the Revolutionary War hero was not known for making public statements against the Catholic Church, and "probably" received last rites from a priest when he died.

There followed a literary duel between the two men in the pages of the *Enquirer*. It moved into general themes: the role of religion in society, whether the Protestant Reformation spawned Democracy, whether the Pope could tell Catholics how to vote, and a variety of other matters. The exchange was picked up by newspapers around the country, and generated something of a national debate via letters to editors from aroused readers.

Morse eventually got the feeling that he was losing the debate—or at least not winning it. The flimsiness of his evidence on Lafayette could not be overcome. The extensive Lafayette-Morse correspondence (of which the bishop was entirely ignorant) obviously offered no recourse. The debate ended when Morse notified the *Enquirer* that, having been "dragged" into a pointless debate, it was time to end this war of words.[12]

Morse did not monitor Catholic activities in the United States, to find any evidence of the "I told you so" variety. A note in the 1840 Letter File lists 300 Catholic churches, ten colleges and thirty-five seminaries in the United States, but the Archive has nothing more in the way of factual information about Catholics. Neither did he become involved in the Catholic schools controversy that raged in Gotham during the 1840s, when Archbishop Hughes demanded public funding for the new Catholic schools, equivalent to the support that was customary for all other schools, most of which, although "public," had some kind of Protestant denominational affiliation. This high-profile controversy dragged on for years, and wasn't resolved until all taxpayer funding for sectarian institutions was ended.

There is a curious coincidence in the fact that a life-size statue of Giuseppe Garibaldi, drawing his sword, stands in Washington Square, New York. It is directly across the street from the art studio and garret where Morse and the Piedmontese envoy launched their Protestant missions project.

Garibaldi's big moment came twenty years later, when he and his Redshirts played a leading role in Piedmont's conquest of the Papal States, en route to the founding of the Kingdom of Italy.

When Gotham's Nativists threw in their lot with the more mainstream Whigs (following Morse's run for mayor in 1826), his reaction was to become a very partisan Democrat. Indeed, the literary jousting match with Bishop Spaulding coincided with a campaign for a seat in Congress he made as the Democratic candidate from his home district in Duchess County, NY. (The Whig candidate won by a narrow margin.) A factor in his choice of Amos Kendall as his business manager, ten years earlier, was Kendall's sterling connections in the Democratic Party. William Cullen Bryant, the famous poet who was also the editor of the country's largest-circulation Democratic newspaper, the *New York Post*,, hobnobbed with the inventor regularly at "the Lunch," in Lower Manhattan, for many years. Local Democratic politicians were regulars at the almost daily coffee house confabs. Bryant often showed up with visiting Democratic luminaries from out of town, one day even bringing the secretary of state, Lewis Cass.

The bid for Congress in 1854 was inspired not only by a desire to curtail Irish Catholic immigration. That issue was rapidly being eclipsed by the growing debate over the extension of slavery into new states, and the growing popularity of abolitionism, especially in the Northeast. Morse viewed the abolitionists as a fundamental threat to the social and political order, no less dangerous than the Irish immigrants. He was convinced, moreover, that the mischievous and secretive hand of a foreign power—Great Britain—was at the root of it all. This needed to be exposed, and there was no better forum for doing this than the Capitol.

Morse took a "long" historical view on slavery that tended to highlight the role of England's ruling class and its wealthy merchants. The British Empire, he noted, was responsible for "forcing it originally on us" in early colonial times. Then, c. 1800, it took the lead in interdicting the trans-Atlantic slave trade, and, more recently, freeing the slaves in its Caribbean sugar islands. The practical effect of all this was to disrupt the economic life of the non-British Western Hemisphere, and thereby open the door to more British influence in that region. Exposure of the sectional divide (North vs. South) in America as a British plot, therefore, "paved [sic] the way for a peaceful and proper abolition of the evil." In other words, understanding the source of the trouble would bring Americans together and make reasonable discourse and compromise possible. (The "abolition" in his letter to brother Sidney referred to abolishing sectionalism, not slavery.) He admitted the possibility—even the eventual likelihood—that slavery would die out eventually, but that was a matter for the slave owners to accomplish on their own.[13]

Sidney agreed with all this, and decided to write a pamphlet about it. Hundreds of copies were run off the printing press at the Observer building, and Morse sent a package of them to Amos Kendall, urging him to hand them out to his friends in Washington. In the early days of the Civil War, Morse did Sidney one better and wrote his own pamphlet, which he self-published under the title, "The Present Attempt to Dissolve the American Union: a British Aristocratic Plot." It was excerpted in the *Journal of Commerce*, and attracted quite a few comments, pro and con, from readers. Among its highlights was the notion that slavery was on the way out, anyway. "The course with slavery," Morse wrote, "is not agitation. I consider it dead already. It has at least got the consumption, and agitation may make it kick..." Some readers found this disingenuous, and the author felt constrained to defend his opinion by claiming, at every opportunity, that he had once made a token $50 donation to a new fund for the purchase and manumission of slaves.[14]

Morse believed that President James K. Polk's annexationist vision in the Mexican War (1846-1848) was bound to provoke and further enliven the abolitionists. All the areas targeted for annexation were considered to have a cotton-friendly climate. "I do not hold to the axiom, My country right or wrong, but My country in the right," he wrote, regarding Polk, who seems to have been the only Democrat Morse ever disapproved of.[15]

Being a religious man, he sought guidance in the Scriptures, and fully believed that he had found therein support for his views. The fact that some well-known New England clergymen of his own congregational denomination were activist abolitionists seemed unimportant and drew no comment. Neither did his own father's support (decades earlier, via his weekly magazine, *The Panoplist*) for Reverend Wilberforce's crusade against the trans-Atlantic slave trade. John Brown, the abolitionist firebrand whom Morse viewed as the devil incarnate, attended Lyman Beecher's Congregational Church in Litchfield, CT. Beecher's daughter, Harriet Beecher Stowe, was the author of the best-selling anti-slavery novella, *Uncle Tom's Cabin*. (This book galvanized antislavery sentiment like nothing else, beginning in 1851 when it first appeared.)

Considering the deep involvement of the Northeastern churches in the debate over slavery, one wonders what brothers Sidney and Richard had to say about it in the *New-York Observer*. By now their newspaper was the region's largest-circulation religious weekly. The *Observer* faithfully reported news items sent in by the various Protestant churches. These sometimes included highlights of sermons on slavery and the reactions thereto. A separate section of the paper was devoted to news and announcements from African-American churches. The *Observer* clearly had a significant readership among black churchgoers. Its position on slavery was to leave all comment to the churches.

Morse followed up his pamphlet on the wirepullers in London with a new one, titled, "My Creed on the Subject of Slavery." Its main point was that "Slavery per se is not a sin." As justification, he pointed out that slavery is not condemned or even lightly criticized anywhere in the Old or New Testaments. The pamphlet went on to explain that slavery is an ages-old social institution, "subject to abuse like any other," and accordingly hedged about by customary and written law.

Among the close friends and relatives with whom Morse exchanged letters regularly, only a few took issue with his view that the Bible was neutral on the subject. One was George Wood, Kendall's right-hand-man at the Washington & New Orleans Telegraph Co. Wood, however, was unable to quote from the Scriptures in support of his position. Instead, he cited the position that had been taken by so many men of the cloth—including Morse's own father, and of course the Reverend Wilberforce. He also quoted John Wesley, the founder of the Wesleyan Church, to the effect that slavery "is the sum of all evil."

Morse agreed with Wood's characterization of his late father, and recalled his own near-meeting with the famous reformer in 1812, when he was a student at the British Academy. But he then pointed out that Jedidiah had never addressed slavery itself—just the slave trade.[16]

Kendall, who still handled all of the inventor's business relating to patent rights and royalties, weighed in once on the religious implications of slavery. He agreed that the Bible offered no clear, direct, or unequivocal moral position, one way or the other. But he then asserted that a "new Christianity" and a "new morality" had taken root, and that he (a regular churchgoer) agreed with it.

Neither Morse nor any of his correspondents ever brought up the topic of what was being preached from Southern pulpits on Sunday mornings, and how it compared with the preaching in Northern churches. The question of what Morse himself listened to on Sunday mornings, at the Presbyterian Church in Poughkeepsie, NY, also never came up. Sidney, who early on seemed so much in tune with his brother's views, thereafter became silent on the subject of slavery and sectionalism in general. Yet the two brothers continued writing to each other with fair regularity. Letters from brother Richard, on any subject, are notably absent from the Correspondence File, and it is worth noting that the *Observer* never made any reference to the Morse pamphlets—not even a third one that appeared late in 1863, on the topic of religion and slavery.

The inventor's position on the sectional controversy, favoring moderation and compromise, was not different from that of his hero, the Marquis de Lafayette. When Lafayette expressed it in a letter to Morse in 1833 (as the Missouri Compromise of 1820 was coming under increasing attack), the artist gladly published his remarks in the newspapers, and the Frenchman thanked him for doing so.

But somehow, Lafayette never got around to giving Morse his opinion on the institution of slavery itself. To a relative in France, however, he once confided: "I would never have drawn my sword in the cause of America if I could have conceived thereby I was founding a land of slavery."[17] Since he was well aware that even George Washington owned slaves, this remark is difficult to credit. One can only assume (assuming that he actually wrote it) that Lafayette was appalled by the huge post-Revolutionary expansion of slavery that was set off by the invention of the cotton gin in 1793.

In private, Morse referred to Abolitionism as "the mother of secessionism...two monsters," and cited Cotton Mather, the great Puritan leader of the 1600s, whom he thought would have condemned Abolitionism as un-Christian. It was, moreover, "the blossom" of Unitarianism, the "progeny of Unitarianism and Infidelity...a hallucination more dreadful in its consequences on human happiness and the order of society than the witchcraft delusions of the 1690s." Abolitionism and Secessionism were "the two atrocious heresies," the products of a "national delirium."[18]

In 1863 he came out with a third pamphlet, "The Ethical Position of Slavery in the Social System," which contained all of his religious arguments.

There was also a racial dimension to Morse's views on slavery, about which he was explicit only in a few letters to George Wood and to some Copperhead (anti-War) Democrat friends. At the outset of the Civil War, he wrote that the "character and necessities of the subject race" debarred any hope for black citizenship. In mid-war, as Lincoln was preparing to emancipate the slaves, he wrote, "Nothing is clearer in my mind than that the status of the African...is that of subjection to the superior race." Morse claimed that black equality would lead to social "degeneracy," marked by the "spectre" of miscegenation. Wood (unlike the Copperhead correspondents) disagreed with all of this, but the verbal jousting between these two distant friends went on and on.[19]

Morse did not have a favorable opinion of the Confederacy, despite his racism. He believed that it was the product of leadership failures and bellicosity on both sides; a passing phenomenon, whose member states (with the possible exception of South Carolina) would eventually find their way back into the Union, in a more-or-less status quo ante relationship.

The selective confiscation of telegraph companies as "enemy property" had an impact on the inventor's income. The criterion for such takeovers was a majority of Northern shareholders. The Confederate Sequestration Act of 1862 cost the Southwestern Telegraph Co. two-thirds of its customer base, and the Washington & New Orleans system was practically wiped out as a private company for the duration of the war.[20]

Morse transferred all of his stock in the two companies to his brother-in-law, Arthur Griswold, who lived in New Orleans and therefore was a Confederate citizen. The aim was to remove the inventor's stock from the category of "alien property," if and when the Confederate government might decide to re-privatize the telegraphs, or compensate the Southerners who owned stock in them.

The fighting in Tennessee and the Lower Mississippi Valley caused extensive damage to the Southwestern's lines. The Washington & New Orleans system, except in Northern Virginia, remained largely intact until Gen. Sherman's March to the Sea in 1864, when retreating Confederates cut the lines. Union military telegraphs everywhere, however, were handled exclusively by the Army Signal Corps, which deployed its own wire.

In March 1861, just weeks before the attack on Fort Sumter, the inventor held an organizing meeting at his house in Manhattan for a Democratic Party anti-war advocacy group. The Society for Promotion of National Unity (SPNU) was the result, and its directors met weekly in the inventor's library.

In 1863 the SPNU was renamed the Society for the Diffusion of Political Knowledge (SDPK). Morse was elected president, and drafted its by-laws. There were real subversives in the SDPK, and Pinkerton investigators (who did the Union's intelligence work under contract) even identified some who were using false identities. People close to the inventor, including his brothers and Amos Kendall, warned of the danger of associating with such people.

Morse came under the scrutiny of the newspapers as well as the Pinkerton detectives. In 1864, a Boston daily mistakenly reported that Morse had been detained and sent to Fort Lafayette for the duration of the war. He believed that his mail was being monitored, complaining at times of open or damaged envelopes.

During the war, drives were organized to send blankets, bandages, etc., to Union prisoners of war. Morse contributed to these, but he also sought opportunities to do the same for Confederates held in Union prisoner-of-war camps.[21] At one point he sent money to a lady in Washington for this purpose, only to learn later that the charity was a fraud.

The inventor-turned-activist contributed $2,000 to the campaign of Gen. George McClellan, the Democratic candidate for president in 1864.[22] He was a member of the general's entourage when McClellan made a final campaign speech from the balcony of the Fifth Avenue Hotel, on election eve. Later the same night, Morse wrote an exuberant letter about the experience to his brother Richard. Richard's reply, a week later, expressed gratification that Lincoln, and not McClellan, had won the election. McClellan did take Gotham (which would soon elect Fernando Wood, a Copperhead, as its next mayor). But Lincoln carried the state and its thirty-five electoral votes, as he had done in 1860.

Ever since completing his second stint as head of the National Academy in 1864, it had been the inventor's custom to turn down requests to serve on the boards of corporations and benevolent societies. He would typically cite the infirmities of old age. In 1869, Morse broke his leg falling down a flight of stairs at home. He was bedridden for two months, and remained wobbly and dependent on a cane thereafter. But the following year the inventor nearly gave in on one offer—the presidency of the African Colonization Society. He began a letter of acceptance, referring to his father's "zeal and benevolence in behalf of the African race," and was gratified that the Society had been founded "by the negroes themselves, under the head of their large hearted and benevolent commander Captain Paul Cuffee." The letter, so remarkably unlike Morse, was unfinished.[23]

Morse the celebrity clearly overshadowed Morse the indefatigable proponent of controversial and "politically incorrect" (even in those days) causes. Public recognition of his inventive exploits even predated the opening of the Washington-Baltimore experimental line in 1844. He was elected a corresponding member of the National Institute for the Promotion of Science in 1841. The following year, the telegraph was a featured exhibit of the American Academy's annual fair, which was held at Niblo's Garden in New York City. People were invited to tap out messages to friends and relatives across the main hall, with the aid of a placard that displayed the Morse code. The *New York Herald* called it "the great invention of the age," and so it was destined to be regarded, with good reason, by people the world over.[24]

In 1848, the first foreign recognition came from an unlikely source—the Ottoman Empire. Sultan Abdul Majid Khan conferred the "Order of Glory" on the inventor. The honor generated some half-serious speculation that the title of "pasha"—an important perquisite of the award—might cost him his U.S. citizenship. (Cyrus Field faced a less fanciful dilemma while being considered for a British Lordship in 1868, following his successful completion of the Atlantic cable.)

Shortly after receiving the Sultan's award, Morse was inducted into the prestigious American Philosophical Society, in Philadelphia. The next year, he was made an honorary member of the American Academy of Arts and Sciences (AAAS), in Boston. His certificate of membership was signed by Asa Gray, the noted Harvard scientist, who was the AAAS's corresponding secretary.

Honors and encomiums also came from societies the inventor had never heard of. The first such was an honorary membership that was conferred by the Philophronean Society, at Hartwich Seminary, in 1851. He sent a cordial acceptance, but was more cautious about later offers.

The first European recognition was the Gold Medal of Wuerttemberg, which was awarded in 1852. It was followed by the Gold Medal of Science and the Arts, from the Austrian Empire (Emperor Francis II) in 1855. Morse became a Knight of the Danebroge, in 1858, by order of King Frederick VII of Denmark.

He was elected a charter member of the new American Geographical & Statistical Society (AG&SS), which was founded in New York in 1855. One of the organizers was Marshall Lefferts, a former promoter of the Bain telegraph who had filed a legal brief against Morse's patent extension in 1854, and was one of the appellants when the case went to the Supreme Court that year. The inventor, whose father was the reputed "father of American Geography," remained active in this organization for the remainder of his life. Board meetings were sometimes held at his house in New York, and he accepted some research assignments, including one that occupied much time during his last trip to Europe, in 1867. His old nemesis, Henry O'Rielly, was also an active member of the AG&SS. (O'R, it should be added, played an important role in organizing Gotham's all-black Union regiment in 1863.)

Morse became a founding member of the American Inventors Society in 1856. He drafted a petition for the group, which was headquartered in New York, requesting the Canadian government to grant automatic recognition to all U.S. patents.

He was also the founding president of the American Asiatic Society (AAS), from 1866 to 1868. Like the AG&SS, it had a strong focus on geography. During his trip to Europe in 1866, he delivered a petition from the AAS to Emperor Napoleon III, urging him to convene an international conference for the purpose of mapping out better trade routes between Europe and the Far East. The Emperor had a longstanding interest in a canal across the Isthmus of Panama. The petition came with a suggested agenda, which included such diverse topics as a study of navigability along the Tigris and Euphrates rivers in Mesopotamia, climate and health issues affecting navigation, and protection against brigands and pirates.

As it happened, Napoleon III was a shareholder in the new Suez Canal Co., a private French enterprise that was just beginning its great project. (In 1869, Empress Eugenie would ride on the inaugural voyage through the canal.) The letter from the AAS reached the Emperor by way of the Foreign Ministry. But the inventor got to meet Napoleon III anyway, being an invitee (along with Mrs. Morse and daughter Lela) at two royal receptions that were held in the Tuileries Palace. This was no surprise: the French ruler had initiated the European Gratuity for the telegraph patent more than a decade earlier.

Morse's next stop on this trip was Geneva, Switzerland, where he was scheduled for a formal induction into the prestigious Societé de Physique et d'Histoire Nationale.

The following year, 1867, Morse was back in Europe, this time as an official U.S. representative to the Paris Exposition. The event showcased new advances in science and technology, and gave awards to scientists and inventors. He received two bronze medals for the telegraph, and attended the official ball for awardees at the Tuileries Palace, dressed in a gold-braided coat, and accompanied by his wife Sarah and son Arthur. Emperors, kings, and other important people were on hand, including the German Chancellor, Otto von Bismarck. Napoleon III, upon seeing Morse in a receiving line now for the third time, paused to comment about the telegraph's steady progress around the world.[25]

Next day, the family Morse was again surrounded by dignitaries during a military review on the Champs Elysee, which they observed from a reviewing stand. Directly across from them were gathered much of Europe's royalty. (A would-be assassin fired a pistol at Czar Alexander II as the event was breaking up.)

After returning to the United States, the inventor spent much time writing his government report on the communications exhibits at the Expo, and the state of European telegraphy in general. The report (which was edited by W. P. Blake, a state department official) was published in 1869 by the U.S. Government Printing Office. It included telegraph maps and copies of documents, obtained from various communications ministries. There was also an account of private interviews with the French Commissioner of Telegraphs, and also with the Duke of Saxony, who had invited Morse to Dresden for discussions on how to improve telegraphic communications in that state.

A welcome home dinner was hosted by William Orton, the former Henry O'Rielly aide who was now president of Western Union, at Delmonico's Restaurant. Cyrus Field, Peter Cooper, Marshall Lefferts, Gen. Erwin McDowell, Ezra Cornell, William Cullen Bryant, and British ambassador Edward Thornton were among the invited guests. Chief Justice of the Supreme Court Salmon P. Chase also attended. Chase had been O'Rielly's counsel during the Frankfort patent case in 1847-1848. As Lincoln's secretary of the treasury, he managed the financing of the Union effort in the Civil War and founded the national banking system.

In 1871 Morse was a founding vice president of the Metropolitan Museum of Art in New York. Some of his pictures were in the original collection. The life-size portrait of his daughter Susan has been prominently displayed at the museum for many years, along with some of his smaller works

On the inventor's birthday, 10 June 1871, a bronze statue with MORSE carved into its granite pedestal was unveiled in Central Park, near the just-opened Met Museum. Commissioned by Western Union, it had been paid for with contributions from the company's employees. The sculptor was Horatio Greenough, designer of the Bunker Hill Monument in the inventor's hometown, Charlestown, MA. (The two had also briefly been roommates in Paris, back in 1831.) William Cullen Bryant was the featured speaker.

(Note: In 1928 the statue was moved to the Hall of Remembrance—later known as the Hall of Fame—at the NYU Heights campus, in the Bronx. When NYU closed the Heights campus decades later, the statue found a new home at the National Museum of American Art.)

The same day, hundreds of Western Union employees enjoyed picnic lunches and boat rides up the Hudson. That night, WU hosted a formal reception for the inventor at the Academy of Music. Telegraphic greetings from all over poured out of a receiver set up on the stage. Each was read aloud to the audience by WU's President Orton. The grand opening of the royal telegraph in Hong Kong had been set for that day. The inaugural message—an obeisance to Queen Victoria—had gone without a hitch. Now Hong Kong replied to the welcoming message that Morse had typed out onstage a bit earlier, and concluded with a birthday greeting. Before reading it, Orton pointed out to the audience that such an exchange previously would have taken the better part of a year.

It was the inventor's last birthday.

NOTES

1. Jedidiah—M, 20 October, 1812.
2. M—Parents, 22 December, 1814.
3. M—Parents, 6 April, 1814.
4. Wilberforce—M, 28 June, 1814.
5. M—Lucretia Morse, 18 June, 1824.
6. Cf. M—Lucretia re the last efforts on behalf of Polish refugees, 11 September, 1832; Lucretia—M, 18 February 1833; M—Lucretia, 5 Marcg 33. The Letter File for 1832—1833 contains all the other correspondence.
7. M—Cooper, 18 July, 1832, 5 October, 1832.
8. M—Parents, 12 December, 1811.
9. Cf. Professor Charles Beck—M, 18 December, 1836; Professor Edward Robinson—M, 18 December, 1836 (from Boston); Silverman, *Electric Man*, p. 127.
10. Morse Archive, Miscellaneous File, 1840.
11. Cf. J. M. Frazer—M, 24 April, 1855.
12. M—*Cincinnati Enquirer*, 26 April, 1855.
13. M—Sidney Morse (brother), 26 January, 1847.
14. M—Sidney, 24 February, 1847; B. Penney—M, 15 November, 1852.
15. M—Sidney, 8 May, 1847.
16. Cf. M—George Wood, 23 February, 1865.
17. Quoted in Gary B. Nash, *The Forgotten Fifth: African Americans in the Age of Revolution*, Cambridge: Harvard University Press, 2006, p. 103.
18. M—Kendall, 1 September, 1861; M—Geo. Wood, 21 April, 1860; M—Charles Mason, 5 April, 1862; M—Richard Morse (brother), 24 September, 1864; M—Sidney Breese (cousin), 10 May, 1863; M—Charles Mason, 20 June, 1861.
19. M—Dr. James Wynne, 14 May, 1861; M—Gideon S. Tucker, 10 April, 1863; Geo. Wood—M, 2 March, 1865.
20. Gen. L. Douglass, Treasurer—M, 24 March, 1862.
21. M—Gen. Hancock, 23 November, 1864.
22. M—J. and P. Tallman, et al., 14 November, 1864.
23. Undated, 1870 Letter Book.
24. Clipping in Letter File, 12 October, 1842.
25. M—Sidney, 27 May, 1867.

Chapter Six

Locust Grove

The early success of the New York, Albany & Buffalo Telegraph Co. (1847–1848) lifted Morse out of poverty. He was fifty-seven years of age, a single man since 1826, the year his wife Lucretia died, and the father of three grown children. As the telegraph industry expanded from that point on, so did his fortune. But he didn't wait for this to happen. As soon as the money began trickling in from the NYA&B, he gave up his hand-to-mouth existence as an art professor, and said goodbye to the little garret room overlooking Washington Square.

Morse took out a $7,500 mortgage on a sprawling 100-acre estate, ninety miles up the Hudson River and just outside the town of Poughkeepsie. A colonial deed in the Duchess County courthouse identified it as "Locust Grove," and so Morse named it. A thousand feet of frontage on the east bank of the great river offered a marvelous view of the Catskill Mountains. It was a place that would have appealed to Thomas Cole and the other Hudson River School painters, and indeed some of them (including Cole and Frederick Church) lived in the Hudson Valley.

One of the mysteries of Morse's art career was his imperviousness to the Hudson River style. It first saw the light of day in the studios and exhibitions of the National Academy of Design, where he served as founding director. But whoever drew Morse's attention to the place certainly knew what they were doing. Contracts were also signed for extensive repairs on the drafty and dilapidated main house.

Built for a large family, the stone structure was a labyrinth of empty rooms. Built for people who loved country living, here was a gregarious individual who made friends with ease, a man who enjoyed the bonhomie and the hurly-burly of life in Lower Manhattan, the most densely populated piece of real estate in all the Americas.

All this was beside the point as far as Locust Grove was concerned. The inventor somehow knew that it would soon be filled with family members, old and new. Such confidence, coming on the heels of twenty years as a live-alone widower, was remarkable. There were no "irons in the fire." His friend, the novelist James Fenimore Cooper, liked to comment, indeed, on his strait-laced ways.

Not long after Lucretia's death, in 1828, he had taken an interest in a girl in Troy, NY, but her parents rebuffed the itinerant and obviously impecunious artist. But then there was the intriguing friendship with Annie Ellsworth, the "What Hath God Wrought" girl who made history by sending the world's first electronic communication, in 1844.

Annie was the daughter of the inventor's former classmate at Yale, Henry Ellsworth, the U. S. Patent Commissioner, who had awarded the patents for the telegraph. In 1843, when Morse spent some weeks in Washington promoting a bill in Congress to finance a proposed experimental line between Washington and Baltimore, he lived in a rental house owned by the Ellsworths. The bill was still pending on the last day of the session, and Morse, in despair, had packed his things and was waiting in the foyer for a lift to the stagecoach station. Annie suddenly burst in, with the happy news: the bill had just passed.

In later years, the inventor would cite this experience as the reason why he chose Annie for the honor that put her name in the history books. But according to Ezra Cornell, the man who actually built the line and then made a fortune in the telegraph business, there was more to it than that. He detailed his impressions in a letter to Samuel I. Prime, Morse's first biographer. The contents never made it into Prime's book, which had been commissioned as an official biography by Sarah Morse, the inventor's widow.[1] The Morse Letter File itself offers some hints of a relationship. In a letter to brother Sidney's wife, Catherine, for instance, the inventor notes "knowing as you do how much I regard her."[2] A letter from George Wood (Kendall's aide at the Washington & New Orleans Telegraph Co.) mentioned that, "Annie was saying how impatient she was that you should get back while the fine weather lasted..."[3] Another letter from Wood notes that, "...she sent her kind regards...and to say that her bouquet had been worn by her to several parties, to the admiration of her friends..."[4] A claim that Morse wrote at least one love poem to Annie has never been verified.[5]

Whatever the truth about their relationship, Annie would never settle at Locust Grove, and seems never to have visited the place. Instead, when her father resigned from the Patent Office, she accompanied him back to their home town, Hartford, CT. The young lady spent the remainder of her life there, a spinster, translating French children's books for publication. The original telegraph tape with the famous message, and an inscription from the inventor, was given to her by the inventor. It ended up as a bequest to the Hartford Athenaeum.

FAMILY, OLD AND NEW...WHERE DID IT COME FROM?

After they lost their mother in 1826, Morse's children continued living with his parents in New Haven. The Walkers in Concord, NH (Lucretia's parents), had set up a modest fund for their upkeep by then. Reverend Jedidiah Morse died the following year, leaving behind a mountain of debt—virtually all of it the result of having co-signed a note for a longtime friend whose bookstore in Boston was facing bankruptcy. His widow found it impossible to cope with the bill collectors, and taking care of three children was also a challenge as age and misfortune began to take their toll on her health. Elizabeth Ann retreated to Utica, NY, her home town, and moved in with her sister and brother-in-law. There was no question of bringing the Morse children along.

The oldest, eight-year-old Susan, was sent off to Concord, NH, and became a ward of the Pickering family (Louis Pickering being a grand uncle, Mrs. Walker's brother). The two boys, Charles and Finn, went to the Richard Morse household in Gotham. They attended Miss Skinner's School together. Finn had been mentally disabled at the age of two by scarlet fever, and although he received generous report cards, including a notation that "he certainly has quickness of mind," his schooling came to an end after two years. Finn then moved on to Utica.

Charles fared poorly at school, and was even put on a limited class schedule. Then he was sent off to a boarding school that charged $5 a week, where he did even worse. It was his custom to send lists of necessities to Richard (usually headed by a request for a rocking chair), but a promise of at least partial fulfillment if he passed all his subjects had no positive result.[6]

Susan went to live with the Richard Morses in 1834. Upon arrival, the almost fifteen-year-old girl penned a letter promising, "I intend father to improve with all my might so that at the expiration of two years I shall have finished my education." Susan proved much more inclined to visit her daddy after classes at Miss Skinner's School, rather than write letters, and indeed she sometimes brought classmates along. There was quite a bit to see—the glass-enclosed photo lab on the roof, the occasional student toiling away in the art studio, and of course that odd-looking contraption sitting on a long table.[7]

True to her word, the young lady graduated at the age of sixteen. She moved back to the Pickerings in Concord, briefly, in 1836, but was quite at loose ends for the next few years. At one point, after a brief sojourn in Baltimore looking for a "situation" (school teacher, governess, companion, or whatever) she wrote to her daddy, "I feel sometimes as if I have no desire whatever to live—life seems without one cheering spot for me...disappointments are the lot of mortals..." Another letter stated, "I miss you and brothers a great deal and feel quite alone in the world." Susan did attract suitors, however. One, named George, interested her, but she could not have him "unless he became a Christian."[8]

Then, at the age of twenty-six (a bit past the prime marrying age for a girl in those days) she met Edward Lind, a young man from St. Thomas in the Virgin Islands. Lind was trying to make his way in the sugar trade, but recent setbacks "depress him at times," she confided to her father. That year, 1846, saw a change of fortune. Lind cleared $9,000, and the two got married. They moved to Puerto Rico, where Lind made a down payment on some sugar acreage and a house near Guyama/Aroyo. Partially to escape her first summer in the tropics, Susan caught a boat to New York in 1847 and spent the summer with her father in his new home on the Hudson.[9]

She did not visit again for three years. In the interim, father and daughter corresponded regularly. Susan began receiving packets of New York newspapers occasionally, after she complained of the isolation in her semi-rural Spanish colonial paradise. At one point, Morse urged her to, "Tell Edward to make money as fast as he honestly can, economize it and bring it here, and turn Yankee." Edward was a much less regular correspondent. In one letter, he noted that Susan had refused to go into town for a travelling variety show. Morse, who had run for mayor of New York in 1826 on a promise to close all the theaters (he came in last of four candidates), replied with congratulations for his daughter. She was too good, he wrote, for "low amusements" and "disgusting exhibitions."[10]

Charles, meanwhile, had learned a trade as a map engraver. He liked to draw, and it was noticed even back in New Haven, when he was a ward of "the Father of American Geography" (Reverend Morse), that he could spend hours copying maps out of the Reverend's geography books. So, when he left school, Morse secured an apprenticeship for him at a New York print shop that did regular business with artists at the National Academy. He then became a freelancer, engraving maps and art work for the newspapers and book publishers.

Charles married Manette Lansing, a girl he met while visiting his Breese relatives in Utica. The event took place there, in 1848. Morse gave the couple $5,000 worth of telegraph stock as a wedding present. He was quite astonished when, barely a year later, Charles came asking for help in paying his rent. His money difficulties grew as Manette gave birth to two sons (Bleecker and Henry) over the next two years. He also launched a lawsuit against Appleton Publishers, claiming non-payment for some maps. Appleton claimed missed deadlines, but agreed to a compromise payment, with a stipulation that the matter was definitively settled. When Charles turned around and filed another suit, Appleton countersued, and won a judgment for

$1,638. Morse ended up paying the judgment. Charles's reputation in the New York publishing world, meanwhile, was tarnished, and jobs became fewer and farther between.[11]

Morse would come to regret his son's choice of a wife. That wedding day proved memorable in more ways than one, however. At the reception, young Finn was sitting alone in a corner when a young lady introduced herself and then escorted him around the crowded room, arm in arm, as if they were a couple. In due course it was the inventor's turn for a greeting, and the young lady introduced herself as Sarah Griswold. She apologized for her slurred speech, quickly explaining that she was deaf, but a good lip reader. Morse got to meet her mother, the widow of an army general, before the festivities were over.

A romance quickly sprang up, but it was not between Morse and the widow Griswold, who was one year younger than himself. Still less was it between Sarah and Finn. No, it was between Sarah (aged 27), and Morse (57). They were married just two months later, in August. The reception took place in the same room, at the Leonard Breese household.

The event set tongues wagging, but only Mrs. Lansing, Manette's mother, openly made an issue of the inventor "robbing the cradle," and rescuing an allegedly helpless young woman from spinsterhood (27 being a late marrying age in those days) by waving his new-found fortune in front of her. Nobody else seemed to care, judging from the attendance at the August event.

Thus began Morse's new family. Sarah moved to Locust Grove, bringing her mother with her. Finn came along, too. A year later, Sarah gave birth to the first of their four children.

The inventor remained busy, first with his patent extension, which triggered legal challenges from the telegraph franchisers, and then with the trans-Atlantic cable project, which would be one of the great challenges of his life. Through it all, Charles and Manette continued with their financial struggles. A new job at a printing plant in Orange, NJ, didn't last. Faced with eviction for non-payment of rent, Charles begged his father for help. Morse's response was exceedingly generous. He bought a house in Brooklyn and gave it to them, after securing a promise that they would pay the annual property tax and keep up with needed repairs. The promise went by the boards when the couple could not come up with the $100 to pay their first tax bill. Son and daughter-in-law were then put on a $100 monthly allowance, but even this did not prevent them from falling deeper into debt.

In 1865, they lost their younger son, Henry, to scarlet fever. Charles then decided to try his luck out west, in the gold and silver mining rush then developing in the Pikes Peak area around Colorado Springs. The idea was to open a business making survey and claim maps. Charles spent four years in Colorado. When he decided to come back home (not having seen his wife at all during that long interim), Morse had to send him the train fare. During his absence, Manette began writing to her brothers in Utica that she and her surviving son, Bleecker, were going hungry and dressed in rags. When Morse got wind of this, he wrote to every one of his relatives, including the Lansings, detailing all the financial assistance she and Charles had been receiving, including the free house and the ongoing monthly allowance. Word soon got around, also, that the inventor would never again be in the same room with his daughter-in-law, and would likewise shun anybody who had anything to do with her.

A psychologist might explain Charles's troubles on a lack of parental support when he was growing up. The Richard Morses had two children of their own, both younger than Charles, and there is no telling just how he fit in.

Morse's second flock of children grew up in a different world entirely. Governesses, tutors, maids, footmen, and the unwavering attention of caring parents, could not produce another clutch of misfits. Or could they? Arthur was born in 1849, daughter Lela in 1851, William ("Willie") 1853, and Edward ("Eddie"), in 1856. Wife Sarah was lucky to have no miscarriages, and all the kids were healthy.

Morse took delight in his new family. In 1854, during a rare absence from home (he was in Washington for a hearing on his patent extension), he wrote to Sarah, "Oh how I miss those dear children and their cheerful voices. When evening comes, I feel the privation of their bedtime prattle, and the opportunity of hearing their lisping prayers."[12] During his absence on the trans-Atlantic project in 1857, he had occasion to write to "my precious wife" again saying how he missed "the little prattlers." Letters ended with fond goodbyes, such as this one: "And goodbye again my dear wife with kisses to all the sweet children from dear Papa."[13]

The children's upbringing would, eventually, present some real challenges. A governess named Sophie Pfisterer was hired in 1859, with the hope that she might impart the rudiments of her native German to the children. Her one-year contract was not renewed, and the family hired a French lady who had been working for the Linds' neighbors, Mr. and Mrs. Thomas Solomon, in Puerto Rico. At the same time, it was decided to send Arthur and Willie to boarding school.

Several months later, the alarming news arrived that Ms. Pfisterer had committed suicide shortly after being admitted to a mental institution in Trenton, NJ. (By curious coincidence, Morse's business manager, Amos Kendall, had just retired from the Washington & New Orleans Telegraph Co. and was now living in Trenton.) Her brother from New York came to Locust Grove to inquire about any problems she might have had while living there. A similar request came by letter from her previous employer, a woman in Augusta, GA, named Ellen Davis. Morse answered both that Ms. Pfisterer had the habit of talking to herself, and wasn't otherwise very communicative.

There the matter ended. Mme. Subit, the new governess, had much better luck and lasted for three years, but with only Lela and little Eddie as her charges.

Home on vacation from boarding school in 1861, Willie accidentally grazed Arthur's head with a bullet from a small pistol he had found hidden in a closet. At that point the boys were enrolled at an academy in Newport, RI. In letters to Morse, after the fall term began, the headmaster described Willie as "a steam engine with legs," and warned that Arthur's foul language and dirty jokes could get him expelled. Willie had been asking to have his pony sent from home, and after the headmaster agreed with Morse that it might have a calming effect on him, his wish was granted. The horse was a recent birthday gift, sent all the way from Puerto Rico by the Solomons.[14]

The next year saw the two boys at school in Pelham, NY, which was only an hour away from home by steamer or railroad. The headmaster reported that Arthur had been caught several times chewing tobacco, and might face disciplinary action. A letter from home ordered the boy to give up the "disgusting and filthy" habit.[15]

Lela was being home-schooled during these years, with the French governess playing a leading role. She also studied piano, and became proficient enough to give occasional recitals at the Presbyterian Church in Poughkeepsie. Her musical endeavors stirred Arthur to take up the violin, which he pursued with a diligence that contrasted sharply with his disinterest in school work. Eventually, during the winter breaks from school, he even accompanied Lela in recitals at the Presbyterian Church in town, and at Morse's Tuesday night at homes in Gotham, after his father bought a second home there.

Neither Arthur, Willie, nor Lela ever went to college. (None of the inventor's first brood ever went to college, either.) It is surprising that Lela did not enroll at Vassar, the new women's college that was founded in nearby Poughkeepsie when she was still a little girl. Morse was one of the founding trustees, and the school very nearly talked him into reviving his career as an art professor. She eventually married a French concert pianist, and spent most of her remaining life in Paris.

Susan Morse Lind had a son named Charles, who graduated from Union College in upstate NY in 1864. He was nineteen—prime military age, and many of his classmates had left school to fight in the Civil War. Charles, a Spanish subject, was one of just a few students at Union who was not a United States citizen. (Puerto Rico, where he was born and raised, did not become a United States territory until 1898.) His parents were on hand for the graduation, and Grandfather Morse—who had paid for Charles's education at Union—treated the Linds and their new Bachelor of Arts degree holder to a leisurely tour of New England. The entire Morse family, including mother-in-law Griswold, went along, and the trip lasted most of the summer.

When it ended, Charles did not accompany his parents back to Puerto Rico. Job opportunities seemed abundant in wartime Gotham, but the young man's interests turned to art, and the studios of the National Academy became a favorite destination. Like his grandpa, he had shown a penchant for drawing at an early age, and everybody thought he had some talent. In 1867—three years out of college—Charles set out for Paris. Funding for this adventure came, of course, from Grandpa. It was an adventure indeed: the young man's first trip to Europe, no prior placement in an apprenticeship, limited experience of living in an unsupervised environment, and little familiarity with French. Morse's judgment in the matter is certainly open to question. One can only speculate that memories of his own life, at that stage, encouraged a good deal of wishful thinking.

During this period, his parents were experiencing serious difficulties. The capture of New Orleans by Union forces in 1862, and the operations of Confederate commerce raiders, caused a sharp decline in sugar exports from Puerto Rico to the United States. Lind's efforts to make new connections with ship operators in New York, where the largest sugar refiners were located, made little progress. He eventually talked Morse into paying to have a schooner built for him, at the New London, CT, shipyard. Work on the boat was interrupted by long delays, however, as New London and all the other shipyards were swamped with Navy orders. It was finally launched early in 1865, but then Lind couldn't find a crew that didn't demand extortionate wages, until the war was over.[16]

Lenders began foreclosure proceedings in 1865. Morse acceded to a loan for $10,000 in "gold dollars," at 6 percent interest and sent along these encouraging words: "Look above, dear Edward, and you will see behind the dark cloud a bright sun." The merchant bank that handled the transaction, Miller & Thebould, warned that the whole amount might be attached or "frozen" immediately upon deposit in Puerto Rico, leaving Lind with no flexibility in handling his debts. That is exactly what happened. Lind even lost the new schooner when it finally arrived from New London later in 1865. But the remittance in "gold dollars" had helped save the family home. Edward and Susan decided to tough it out in Guyama/Arroyo.[17]

The Linds were taken by surprise when Morse offered them an opportunity to visit their art student son in Paris. The inventor was planning a mixed, year-long business and pleasure trip to Germany, France and Switzerland, beginning in summer 1868. Among the participants in the 1864 perambulation of New England, only Finn was definitely out of the picture. He would be staying with Mary Davis, the inventor's cousin in Utica. Mother-in-law Griswold would also be missing during the early going, while she visited her brother Thomas in New Orleans.

Arthur and Willie would be taking time off from school, but Morse found somebody at church who was willing to accompany them as a tutor/guardian, with the idea that they would sometimes go off on their own. The Morse coachman also went along, as a valet, and so did the family maid.

The two boys and their tutor, Colonel John R. Leslie (a bachelor who lived with his two sisters in Poughkeepsie) were indeed separated from the others most of the time. They met up with the others in Geneva, early on, and Morse looked on happily as all four of his youngsters cavorted in the snow at the foot of Mont Blanc. The two older boys and Colonel Leslie then disappeared for an extended period, for parts unknown in Germany. Edward Lind returned to Guyama.

Morse and the others (including Susan Lind) spent most of the winter of 1868–1869 in Dresden, capital of the Duchy of Saxony. He conferred with the Saxon minister of telegraphs on matters relating to ventures in nearby Eastern Europe. (Although Austria-Hungary and Russia had each contributed 70,000 francs to the 400,000 francs gratuity to Morse back in 1854, telegraph poles were still a rare sight in large areas of their domains.)

In February the family moved to Berlin. There, the inventor had meetings with Werner Siemens, co-founder (with his brother) and head of Siemens & Halske. That company was Europe's largest manufacturer of telegraph equipment. It also had a research and development department that had acquired some patents for improvements. Of particular interest was the prospect of doing business with Western Union, and building another trans-Atlantic cable, to join the one that the Cyrus Field interests had completed in 1866.[18]

Paris was the final destination—minus Arthur, Willie, and the Colonel, who were still in Germany. Morse rented the entire floor of an apartment building, and "a cellar room for our wine," one block from the Champs Elysee. Grandson Charles had a reunion with his mama; and his sponsor, Grandpa Morse, got to see what he had accomplished over the past several months in his new surroundings. The young man then astonished everyone by going home to Puerto Rico.

The family returned home in May. Arthur and the Colonel stayed behind for a summer of travel through Scandinavia. When they got back, Morse found the Colonel a job at Western Union's headquarters on lower Broadway in Manhattan. Young Arthur, his education over, found work at at Lockwood & Co., Wall Street stock brokerage.

After just a few months in his first job, Arthur fled the country, catching a ship bound for San Francisco, but getting off at Valparaiso, Chile. Exactly how he got into trouble is unclear. Morse wrote to brother Sidney only in vague terms, referring to "…a domestic trouble preying on my spirits respecting Arthur, the details of which are only for the private ear." Willie, who was now attending Phillips Academy in Andover, MA, received a cautionary letter, citing his older brother as an exemplar of the "privations" and "sorrow" that attended upon misbehavior.[19] Whatever Arthur's problems were, he returned in 1870, and promptly went to work as a bookkeeper at the Western Union headquarters, which was only a short walk from his former employer.

Willie at this point was nearing the end of his formal education. Disciplinary problems continued unabated. Among the litany of complaints sent home by Mr. Taylor, the headmaster at Phillips, was the alarming news that the teenager spent most of his free time hanging out in "public saloons." As the holidays approached, Willie got into really hot water by insulting the headmaster in front of other students. Faced with an alternative of expulsion or a public apology, the lad opted for the latter, but only after receiving a frantic letter from his father. Willie was not allowed to come home for the holidays.

The next term was his last. In June 1870, shortly after it was over, the rambunctious lad also said goodbye to Locust Grove. Nearly two weeks later, a Union Pacific train delivered him to his destination: Dodge City, KS. For this well-travelled son of privilege, even a brief visit to such a godforsaken hellhole seemed incongruous. But then young Willie did love horses, adventure, and saloons. Dodge had plenty of all three. For Willie, it was the place to be, and he stayed.

Fears that he was throwing his life away—at least in the literal sense—would prove unfounded. Sharing campfire beans with hard-bitten desperadoes and herding longhorn cattle to the Dodge and Abilene railheads suited him just fine. Years later, when the aches and pains of the strenuous life began to catch up with him, Willie found work as a cowboy troupe performer in Buffalo Bill's Wild West.

By his own lights, Willie's life was purposeful and rewarding. None of his forebears among the Morse offspring could say as much. Half-brother Finn, of course, had little choice in the matter. His later years were spent among Utica relatives with whom he had formed an attachment, and who collected $50 a week from Morse for his upkeep.

Willie's younger sister, Lela, married a French concert pianist and moved to Paris. She often accompanied her husband on tours. So, for one product of Locust Grove, life was lived among the *haute monde* of Europe; for another, it was flea-bitten livestock, rattlesnakes, and belly-up-to-the-bar. Eddie, the last of the Morse brood, was the only one to attend college, for just one semester. He alone appears to have taken his father for a role model. Eddie became an accomplished artist, and enjoyed a modest success painting portraits of Gotham's elite, well into the twentieth century.

To the very end of his life, Morse carried the burden of son Charles and family's heavily subsidized existence. One wonders if he could afford all this without hardship—especially after the disappearing $10,000 loan to Edward

Lind in Puerto Rico. Annual income reports to the Duchess County tax assessor indicated that throughout this period (since about 1860) the inventor's annual income averaged between $30,000 and $35,000. It came principally from Western Union's regular annual 6 percent dividends. WU shares made up about four-fifths of his wealth. His other major investment was the Empire Mutual Life Insurance Co., and indeed he sat on the company's board of directors. (Ex-President Millard Fillmore was another board member). Morse also held shares in the Morse Insurance Co. (which gave him the shares in return for the use of his name and image). Locust Grove, bought for $7500 in 1849, was worth at least twice as much since major improvements had been completed. These included a doubling of living space in the stone house, a refurbished apple orchard, and several acres of grape vines that Morse introduced during the 1850s, using rootstock from Chile that was new to the New York State grape growing industry. The vines produced table grapes only, even though the Morses were accustomed to having wine with their meals. There was also a resident farmer, who paid rent and grew assorted fresh produce, including cool climate items, like turnips and kohlrabi, that kept the growing season going from March until early November.

The inventor's brothers, Richard and Sidney, invested most of their earnings from the *New York Observer* in Manhattan real estate. Their first venture was a property on Broad Street (just off Wall Street), where they built a three story office building—renting out two floors, and keeping one for their newspaper. Their state-of-the-art Hoe printing press was in the basement, and it earned additional revenue doing outside work. Years later, they teamed with their older brother and purchased four connected rental properties on West 22nd Street, a block west of Broadway. Morse decided to make one of them—a three story building undergoing extensive repairs—his second home, mainly for relief from the rigorous Hudson Valley winters.[20]

Morse also maintained a slush fund for risky investments, which invariably were in mining ventures. He invested in the country's first oil boom, at Titusville, PA, and in the Colorado Mining Co., when Charles was making survey maps for the company in Colorado Springs. Morse also sank money into the Sierra Madre Gold Mining Co., the Keystone Copper Mine, and some Virginia City, NV, ventures during the silver rush there. All of these gambles lost money; Titusville cost him $3,000, and toward the end of his life he estimated the western mining losses at $19,000 total. All this amounted to little more than an expensive gambling hobby.[21]

It would seem, then, that the steady drain of funds to son Charles, to son-in-law Edward Lind, and assorted other family members, though surely much regretted, was never a threat to the inventor's finances. Morse, in fact, was donating to various causes throughout these years. He contributed regularly to theological seminaries, and gave annually to his alma mater, Yale University. At one point, he pledged $10,000 to provide a new building for the school of theology, on the condition that it be named after his father. (Jedidiah earned his D.D. at Yale.) School officials told him that an additional $30,000 in outside money would have to be raised. Morse tried, unsuccessfully, to round it up on his own, and the project was then dropped. The inventor also gave $3,000 to NYU (where he invented the telegraph and taught art as a member of the charter faculty), to help the school get over a serious cash crisis during the 1850s.

Sarah Morse had her own priorities when it came to charitable giving, and her husband supported them enthusiastically. One was primary education for deaf children. She had a firsthand knowledge of the challenges that faced young people with a hearing impairment, and a school for the deaf in Washington was especially favored with regular annual donations. Her other priority was women's rights, and especially women's access to higher education. She was a friend of Emily Howland and other pioneering activists in this area. The inventor's very important role in the founding and early development of Vassar College, in nearby Poughkeepsie, owed much to his wife's influence. Rutgers Women's College in NJ was also a regular recipient of cash gifts. (Every Christmas, the Morses received a student-made present from Rutgers.) Western Union also increased its hiring of women as a direct result of the inventor's urgings, during the regime of William Orton, the company's second president. The two men had been acquainted since 1845.[22]

The inventor would never have the satisfaction of seeing Charles stand on his own. In 1870, an especially dreadful situation unfolded. His current employer, Lippiatt Silver Plate & Engraving Co., offered Morse the opportunity of becoming a partner. He accepted, mainly because it would tend to provide some job security for his son. Lippiatt thereupon went bankrupt, and the other two partners had no money in the bank. As an unlimited partner, Morse was stuck with the entire debt. J. O. Lindsay, a Wall Streeter with some experience in wealth management, offered his services, saying he could save the inventor some money. He presented some impressive letters of recommendation from happy clients. Morse was mightily impressed with his self-confidence and knowledge of finance. When Lindsay asked for $25,000, with the assurance that it would restore the business to solvency, it was handed over without question.

Months went by, with no news of progress. These dark days were made even darker by a fall down a flight of stairs at home, which left the inventor bedridden for two months. Friendly inquiries produced only a request for $3,000 more, which was not sent. The tone of Morse's letters changed markedly, and finally, after being accused of "gross swindling" and threatened with legal action, the slippery adviser revealed that all the creditors had signed off on their claims, for a total of just $13,000. Most of the balance, however, had been used for "expenses," leaving barely enough to cover the adviser's fee. The inventor, eighty years of age, was still recovering from a bad fall down on a staircase at Locust Grove. At one point, during the tense months of this imbroglio, he confided to his attorney that the matter was "seriously affecting my health." In fact he had only months to live. During the interim, Lippiatt was put on the block, and sold for barely $5,000. The new owners, as a condition of the sale, signed an irrevocable employment contract with Charles. It guaranteed him nine years employment at an annual salary of $2500. [23]

Morse breathed his last breath in May 1872. His younger brothers had preceded him to Green-Wood Cemetery in Brooklyn, where he owned a family-sized plot atop a steep-sided knoll. Richard, who was two years the inventor's junior, had died suddenly while vacationing in the Bavarian Alps in 1868. Sidney, the youngest, had passed away just the previous January.

A three-sided obelisk that Morse designed dominates the Green-Wood site. It somewhat resembles the towering Bunker Hill monument in Charlestown, MA, the brothers' hometown. Horatio Greenough, who built the Bunker Hill memorial, briefly roomed with Morse in 1832, when both happened to be in Paris. It was during that sojourn that Morse took some time off from his Louvre painting project, and had his historic encounter with the fugitive inventor, Harrison Gray Dyar.

In June, Arthur was visiting his uncle Griswold in New Orleans, and slipped off a platform between the cars on a local train. His mutilated body was shipped home for burial.

Susan's son Charles, who never left Puerto Rico after abandoning his art studies in Paris, shared the family struggles in the sugar export business. Susan continued to be plagued by bouts of dengue fever, and life with Edward Lind had taken a decided turn for the worse, when she found out he had been cheating on her. They were no longer on speaking terms, and even had a partition bisecting their dining room table.

In 1880, Charles blew his brains out with a pistol shot. His father passed away three years later. In 1885, Sarah Morse, the inventor's widow, invited Susan to live with her. Susan accepted, and boarded a ship for New York.

Morse had once written of Locust Grove: "The country is in all its glory, the trees in their richest green, the flowers in their gayest dress and the birds singing their sweetest notes."[24] Now Sarah received a telegram: Susan had disappeared during the voyage, and was presumed lost at sea.

NOTES

1. Cornell—Prime, 28 April, 1873.
2. M—Catherine, April, 1844 (no day given).
3. Uncatalogued folder in Letter File.
4. Wood—M, 21 July, 1844.

5. Morse Letters and Journals, Vol. 2, p. 217–218; Vail Telegraph Collection, Archives, Smithsonian Institution.
6. Charles—M, 9 October, 1842.
7. Susan—M, 31 October 1834.
8. Ibid., 8 May, 1818; June, 1814; July, 1841.
9. Ibid., 22 May, 1845; M—Sidney, 24 February, 1847.
10. M—Susan, 26 May, 1847.
11. M—W. E. Smith, 30 March, 1865; G. D. Lord,—M, 23 November, 1867.
12. M—Sarah, 17 February, 1854.
13. E.g., M—Sarah Morse, 6 June, 17 August, 1857.
14. M—Arthur, 31 May, 1862.
15. Ibid., 7 April, 1867.
16. Cf. M—Huntington & Co., 21 April, 1864.
17. Cf. M—Lind, 5 May, 1865; 3 October, 1867.
18. Siemens had patents for his improvements in Prussia, but Morse had none there for his original invention.
19. M—Sidney, 14 August, 1869; M—Willie, 7 October, 1869.
20. Richard—M, 30 May, 1849.
21. Tax Return to Thomas Jaycox (Duchess County), 10 February, 1869.
22. Cf. M—Howland, 23 October, 1861.
23. M—J. P. Lindsay, 22 January, 1807 February, 1872; M—John E. Parsons, 12 February, 1872; M—Mrs. M. Goodrich (cousin), 17 February, 1872; M— Parsons, 12 March, 1872.
24. M—Arthur, 31 May, 1862.

Concluding Remarks

RIGHT TIME, RIGHT PLACE, AND SAMUEL F. B. MORSE

On the topic of Right Time: Some inventions come out of nowhere and catch people by surprise. The telegraph was not of that type. It belonged to that category of anticipated inventions, the anticipation raised by new scientific discoveries and experiments. Morse personally experienced "the state of the art" of electro-magnetic research, c. 1810, in Professor Silliman's science class at Yale. A wire, one end stuck in a battery, passing down a line of students, made his hand tingle. By the 1820s, the science was such that people were beginning to think seriously of practical applications. Electronic communications headed everybody's list. We have seen that many alternative telegraphs were under development, contemporaneous with Morse's, during the 1830s. It seems safe to say, then, that by about 1850, even if Morse had never existed, some sort of telegraph(s) would have been in operation, and the Electric Age would have been underway.

An afterthought on Right Time: The telegraph was an exponential advance, nothing even remotely like it having ever existed before. Everything that came after was incremental to it. If it were not the Morse Telegraph (which sent all of its competitors to the scrap heap) how would this have affected the unfolding of the Electric Age? This is an interesting question that historians of technology may wish to consider in greater depth, because it has implications for the unfolding of all new technologies.

What about the Right Place? What place was that? Gotham? The City University on Washington Square? The university provided Professors Leonard Gale and John Draper, student Alfred Vail, an inexpensive work space, a room to live in, a very modest living, and (in 1837) a venue for the successful public debut of his telegraph. The academic affiliation also made the distinguished scientist Joseph Henry accessible for advice on issues of electro-magnetism.

Location-wise, the university was the epicenter of this profound revolution. Within a narrow geographical radius there was Yale University, where the undergraduate Morse first became acquainted with electricity; Rensselaer Polytechnic and Princeton universities, where Joseph Henry did his pioneering experiments; and just uptown, Columbia University, where Professor Dana's Athenaeum lectures revived Morse's interest in electro-magnetism and (via subsequent visits to Dana's lab) kept him current on that new and dynamic field of scientific research. His interest in the subject was certainly no secret, and it explains why his brother-in-law took the trouble of arranging the meeting with Dyar in Paris. (Chapter 4 tells that story in detail.)

The New York, Albany & Buffalo Telegraph Co. (founded in 1847) also falls within this range. The NYA&B proved that the telegraph was a good deal more than an expensive luxury (as the telephone was during its early decades). As the first successful enterprise in the Age of Electricity, it proved that investors could do much worse than buy stock in a telegraph start-up—whether it be in Cincinnati, St. Louis, New Orleans, or San Francisco.

The NYA&B's network included the hometown of Henry O'Rielly and the Rochester Group (the latter being founders of Western Union); Utica, its corporate headquarters, and the home of numerous relatives of Elizabeth Ann Breese Morse, the inventor's mother; and Ithaca, hometown of Ezra Cornell, whose many important contributions to the early telegraph industry cannot be exaggerated. The centrality of the Empire State, and Gotham in particular, is a notable feature of the Electric Age that persisted well into the twentieth century (via Edison and others). It deserves further study by economic geographers.

Bibliography

Documents:

Cyrus Field Archive. New York Public Library.
Samuel F. B. Morse, Archive and Letter Files. Library of Congress. Washington, D.C.
Henry O'Reilly Archive. New-York Historical Society.
Francis O. J. Smith Archive. New York Public Library.
Western Union Archive. New York Public Library.

Morse the Artist:

Kloss, William. *Samuel F. B. Morse*. New York: Harry N. Abrams. 1988.
Larkin, Oliver W. *Samuel F. B. Morse and American Democratic Art*. Boston: Little Brown. 1954.
Metropolitan Museum (New York) *Samuel F. B. Morse, American Painter*. Catalog for 1932 show.
National Academy of Design. *Samuel F. B. Morse, Educator and Champion of the Arts in America*. New York: National Academy of Design. 1982.
Staiti, Paul J. *Samuel F. B. Morse*. New York: Cambridge University Press. 1989.
———. "Samuel F. B. Morse in Charleston, 1818–1821." *South Carolina Historical Magazine*. Vol. 79, No. 2. p. 87–112. 1978.

General Biographies:

Mabee, Carleton. *The American Leonardo: a Life of Samuel F. B. Morse*. New York: Knopf. 1943.
Prime, Irenaeus. *Life of Samuel F. B. Morse*. New York: Appleton Publishers. 1875.
Quackenbush, Robert M. *Quick, Annie, Give Me a Catchy Line: a Story of Samuel F. B. Morse* (City?) Baker & Taylor. 1983.
Silverman, Kenneth. *Electric Man: the Accursed Life of Samuel F. B. Morse* New York: Knopf. 2004.

Invention of the Telegraph:

Anonymous. *Getting Right the Invention of the American Electro-Magnetic Telegraph*. Washington, D.C. 1871.
Americans in Paris. *Report of the Dinner by the Americans in Paris, August the 17th, at the 'Trois Freres," to Prof. Samuel F. B. Morse, In Honor of the Invention of the Telegraph, and on the Occasion of Its Completion under the Atlantic Ocean*. Paris: E. Briere. 1857.
Cole, Donald B. *A Jackson Man: Amos Kendall and the Rise of American Democracy*. Baton Rouge: Louisiana State University Press. 2004.
Curtis, Benjamin Robbins. *Argument of B. R. Curtis, Esq., of Boston, in the Case of Francis O. J. Smith, Complainant, Vs. Hugh Downing and Als. Respondents, in the Circuit Court of the U.S., Mass. Dist., Hon. Levi Woodbury, Judge, Presiding, for the Infringement of the Letters Patent of Samuel F. B. Morse for the Electro-Magnetic Telegraph, June 25 and 26, 1850*. Portland, Maine: F. W. Nichols. 1850.
Denslow, Van Buren. *Thomas A. Edison and Samuel F. B. Morse*. London: Cassell. 1887
Dorf, Philip. *The Builder: a Biography of Ezra Cornell*. New York: Macmillan 1952
Gifford, George. *Argument by George Gifford, Esq., of NY, Delivered in Dec. 1852, Wash., before the Supreme Ct of the U.S., in the Case of Henry O'Reilly, Et Al., at Appellants*. New York: William C. Bryan. 1853.
Hayes, Thomas M. *The Opening and Closing Arguments of Hon. Thomas M. Hayes and Hon. Francis O. J. Smith, before the Superior Court of Suffolk County, Mass*. Portland, Maine: Brown, Thurston & Co. (Motion for a Craig retrial).
Kendall, Amos. *Autobiography of Amos Kendall*. Boston & New York: Lee & Shepard. 1871.
O'Rielly, Henry. *Exposure of the Schemes for Nullifying the "O'Rielly Contract" for Extending the Telegraph between the Atlantic, Lakes and the Mississippi*. St. Louis. 1847.
O'Rielly, Henry. *Henry O'Rielly & Others, Appellants, Other Respondents Vs. Samuel F. B. Morse & AL*. New York: William C. Bryant. 1852.
Smith, Francis O. J. *Defendants Proofs. Depositions by Cornell, et al., in Morse v. Smith suit, 1850*. Boston. 1851.
Turnbull, Lawrence. *History of the Invention of the Electric Telegraph*. New York: William C. Bryant. 1852.
Vail, Alfred. *The Decision of the Great Telegraph Suit of Samuel F. B. Morse and Alfred Vail v. Francis O. J. Smith*. New York: Ira Berry, Printers. 1857.

Telegraph History

Briggs, Charles F. *The Story of the Telegraph, and a History of the Great Atlantic Cable*. New York: Rudd & Carleton. 1857.
Brown, Charles H. *First Telegraph Line Across the Continent: Charles Brown's 1861 Diary*. (Editors: Mihelich, Dennis H., and James E. Potter). Lincoln, Neb.: Nebraska State Historical Society Books. 2011.
Cardello, Mario J. *Utica: Its People and Events*. Utica, NY: Observer-Dispatch, 2007.
Coe, Lewis. *The Telegraph: a History of Morse's Invention and its Predecessors*. Charlotte: Baker & Taylor, 1993.
Cooke, Thomas Fothergill. *The Charge Against Charles Wheatstone of "Tampering With the Press," as Evidenced by a Letter of the Editor of the Quarterly Review in 1855*. Reprinted from the *Scientific Review*. London: Bath, Simkin, Marshall & Co. 1868.
Dickerson, Edward N. *Joseph Henry and the Magnetic Telegraph*. New York: C. Scribners Sons. 1884 (Princeton speech).
Fahie, J. J. *A History of the Electric Telegraph to the Year 1837*. New York: E. & F. N. Spon. 1883.
Field, Harry M. *History of the Atlantic Telegraph*. New York: Scribner 1868.
Jeans, Wm. T. *Lives of the Electricians: Profs. Tyndall, Wheatstone & Morse*. London: Whittaker & Co. 1886.
Hubbard, Geoffrey. *Cooke and Wheatstone and the Invention of the Electric Telegraph*. London: Routledge & Kegan Paul. 1965.

Kieve, Jeffrey. *The Electric Telegraph: a Social and Economic History*. Newton Abbott, U.K.: David & Charles, Publishers. 1973.
Lubrano, Annteresa. *The Telegraph: How Technology Innovation Caused Social Change*. New York: Garland Publishers. 1997.
Marland, E. A. *Early Electrical Communication*. London: Abelard-Schuman. 1964.
McCormick, Anita. *The Invention of the Telegraph and Telephone in American History*. Charlotte: Baker & Taylor, 2004.
Prescott, George B. *History, Theory and Practice of the Electric Telegraph*. Boston: Tichnor & Fields, 1859.
Reid, James D. *The Telegraph in America*. New York: J. Pohlhemus, Printer, 1885.
Schierr, George (ed.) *The Electric Telegraph: a Historical Anthology*. New York: Arno Press. 1977.
Taylor, William B. *A Historical Sketch of Henry's Contribution to the Electro-Magnetic Telegraph, with an Account of the Origin and Development of Morse's Invention*. Washington: GPO (Smithsonian Institution Publications). 1878.
Thompson, Robert L. *Wiring a Continent: a History of the Telegraph Industry in the United States: 1832–1866*. Princeton: Princeton University Press. 1947.
Towers, Walter K. *Masters of Space*. New York: Harper Bros. 1917. (Note: This book is about Morse, Bell and Marconi.)
Vail, Alfred. *The American Electro-Magnetic Telegraph: with the Report to Congress*. Philadelphia.: Lea & Blanchard. 1844.
Vail, Alfred. *Early History of the Electro-Magnetic Telegraph, from the Letters* and *Journals of Alfred Vail*. New York: Hine Brothers. 1914.
Williams, Albert N. *"What Hath God Wrought." May 24, 1844*. Princeton: Princeton University Press. 1944.

Morse's Private Life

Davidson, Lucretia Maria. *Amir Khan, & Other Poems*. (Introduction by "Professor Morse.") New York: G.C. & H. Carvill. 1828.

Index

Adams, John Quincy, 15–16
Albany & Buffalo Telegraph Co., 44, 45, 48
Alexander II, Czar, 17, 53
Alexis, Archduke, 17
Allston, Washington, 2, 3, 11, 16, 17
Alston, William, 6
alternative telegraphs of 1840s, 56–58. *See also* Morse Telegraph; telegraph system
American Academy of Fine Art, 8
American Telegraph Co., 55, 58, 61, 72
American Telegraph Convention, 61
American Telegraph Federation, 61
Ann, Elizabeth, 111
Arago, M. Francois, 30, 31, 67, 78
Arthur, 117
Atlantic, Lake & Mississippi Telegraph Co., 42, 50
Atlantic & Ohio Telegraph Co., 44

Bain, Alexander, 59–60
Ball, Caroline, 6
Baltimore & Ohio Railroad, 36
bathometer, 56
Beaumont, Elie de, 84
binary or digital communications, 26
Blake, William P., 77
Boston-Halifax line, 45
Boston-New York line, 45, 48
Brown, John, 101
Bryant, William Cullen, 9, 107

Buchanan, James, 54
Bunker Hill Monument, 12, 107, 121
Burbank, David, 40
Buren, Martin Van, 27

California Telegraph Co., 61, 62
Campbell, Sir John, 29
Case, Eliphalet, 43, 45, 48
Catholic schools controversy, 99
Chappe telegraph, 30
Chase, Salmon P., 85, 107
Church, Frederic, 12
Clarke, George, 13
Clausing, Henniger, 97
Coffin, Isaac N., 32
Cohen, David, 12
Cole, Thomas, 12, 16, 109
Colt, Samuel, 33
Columbia Telegraph, 51
commissions, 5; Eli Whitney 8; of General William Ledyard, 10; of local judges, 10; from local plantation owners, 6; Mrs. Caroline Ball, 6; panoramic scene of U.S. House Chamber, 7; of "Prince Arthur", 10; small and simple head-and-torso compositions, 6; of U.S. House of Representatives, 6, 8; of William Alston and his family, 6
Confederate Sequestration Act of 1862, 103

Index

Cooke, William Fothergill, 67, 80. *See also* Wheatstone-Cooke telegraph
Cooper, James Fenimore, 9, 13, 15, 18, 95, 96, 110
Cooper, Peter, 53, 107
Cornell, Erza, 35, 36, 37, 44, 51, 84, 107, 110
Craig, David H., 45

Daguerre, Louis, 14
Dana, John Freeman, 77, 87, 88, 124
Dana, R.M., 88
Daniell, J. F., 24
Davy, Sir Humphrey, 23, 29, 67, 86
Doolittle, Thomas, 7
Draper, John W., 14, 24, 68
Durand, Asher B., 12, 16
Dwight, Dr., 2
Dyar, Harrison Gray, 78–81, 89; battery-related experiments, 78; challenging Morse's patent, 78–80; desire for anonymity, 81; "Interrogatories", 79; in Philadelphia, 81. *See also* Morse Telegraph; telegraph system
Dyar, Samuel B., 80
Dyar language, 81

electricity, phenomenon of, 22
Elgin, Lord, 31
Ellsworth, Annie, 34, 37, 110
Ellsworth, Henry, 34, 37, 110
Empire Mutual Life Insurance Co., 118

Faraday, Michael, 54, 67
Faxton, Henry O., 48, 49
Field, Cyrus, 52, 53, 55, 58, 107, 117
Fillmore, Millard, 34
Finley, E. Samuel, 2
Finley, James, 5
Fisher, John, 35–36, 36
Fulton, Robert, 1

Gabersham, R. S., 79
Gale, Leonard, 24, 32, 47, 48, 68–69, 83; early years in Washington, 69; as "patent consultant", 69; presence in the Patent Office, 69. *See also* Morse Telegraph; telegraph system
Glass, Elliott & Co., 55

Greenough, Horatio, 12
Griswold, Arthur, 103

Harrower, Rev. Charles S., 81
Henry, Joseph, 24, 68, 71, 73–78, 88; 1830-31 bell ringer experiments, 74, 75; electric doorbell, 24; experiments in electro-magnetism, 74; Frankfort decision in *The Telegraph Companion*, 76; Morse *vs.*, 74–77. *See also* Morse Telegraph; telegraph system
Hone, Philip, 23
Huntington, Daniel, 16

Inman, David, 16

Jackson, Charles, 13, 22–23, 49, 82–88, 89; claim as co-inventor with Morse, 83–84, 85–88; Morse et. al *vs.* O'Rielly et al, assertions in, 85. *See also* Morse Telegraph; telegraph system
Johnson, Cave, 34

Kendall, Amos, 40, 48, 58, 61–62, 70, 102, 104
Khan, Sultan Abdul Majid, 105
Kloss, William, 4

Lafayette, Marquis de, 10, 15, 95, 102
Lansing, Manette, 112
Lee, George W., 12
Lefferts, Marshall, 107
Leslie, Charles, 3
Leslie, Colonel John R., 117
Lind, Edward, 112, 116, 117, 118, 119
Lindsay, J. O., 120
Lippiatt Silver Plate & Engraving Co., 120
Locust Grove, 15, 18, 52, 69, 81, 110–121
Louis XVIII, King, 94
Louvre Museum, 12

Magnetic Telegraph Co., 40, 44, 55, 60, 61
McClellan, Gen. George, 104
McDowell, Gen. Erwin, 107
Metropolitan Museum of Art, 18, 21, 107
Meyerhoff, Baron Peter, 31
Monon's semaphore, 32, 33
Monroe, James, 5–6, 94
Morse, Charles, 111, 112, 115–116, 121

Morse, Elizabeth Ann, 1
Morse, Finn, 111, 113
Morse, Jedidiah, 1, 4, 93, 94, 102, 111; *The American Gazetteer*, 2; *American Universal Geography*, 2; financial crisis, 7–8
Morse, Lela, 115
Morse, Richard, 3, 4, 8, 11, 83, 101, 104, 111, 114, 119, 121
Morse, Samuel Finley Breese, 1, 44; on abolitionism, 100, 103; "Amr Khan and Other Poems" (book), 11; anti-Catholic adventures, 98–99; anti-Catholic ardor of, 97; in boards of corporations and benevolent societies, 104, 105, 105–106; clay modeling, 4; as clerk in Mexico, 95; commissions, 5, 6; family, 111–113; finance from Academy, 10; first art lessons, 1; idea of miniaturizing masterworks, 13; interest in politics and public affairs, 93, 95, 100, 104; and issues with O'Rielly, 41–44; lectures on art, 10; new family, 113–115; observations about war, 93–94; photographs, 15; portraits, 5; Protestant missionary project in Italy, 98; reasons for quitting art, 15–18; recognition, 105; on refugee problem and immigration, 95–97; risky investments, 119; in Rome, 11–12; schooling, 2; on sectionalism, 102, 103; on slavery, 94, 100, 101, 102, 103; studio at NYU, 14; as teacher at NYU, 14; "the dying Hercules" painting, 4; view on slavery, 94, 100; as a writer, 11. *See also* Morse Telegraph
Morse, Sarah, 120
Morse, Sidney, 3, 4, 8, 11, 83, 100, 101, 119
Morse, Susan, 1, 7, 13, 14, 21, 107, 112
Morse code, 25–26; dot-dash system, 26
Morse Insurance Co., 118
Morse Telegraph, 123; contract with St. Germain Railway Co., 30; demos, 27, 30; efforts to secure royalties, 47; English patent, 31; Franklin Institute's official report on Morse's demonstration, 28; marketing of, 27; Morse's objections to patent to House telegraph, 56–58; question of origins and originality, 67–68; Smith's offer to, 28–29; U. S. patent, 28; use of alphabets, 57
Morse, Thorwaldsen, 17
Morse, Willie, 117, 118
Morton, William, 87

Napoleon III, 106
National Academy of Design, 1
National Academy of the Arts of Design, 9; annual show, 1831, 12; expansion of buildings, 16; fund drive, 16; Morse's contribution to art library, 12; rise of, 9
New Orleans & Ohio Telegraph Co., 51
New York, Albany & Buffalo Telegraph Co., 48–49, 51, 62, 109, 124; growth, 50; network, 124
New York, Newfoundland & London Telegraph Co., 53
New York and Mississippi Valley Telegraph Co., 60
Nicholas I, Czar, 31
North American Life Insurance Co., 72
NY, Newfoundland, 56
NYU studio, 1

O'Rielly, Henry, 41–44, 49, 56, 64, 69, 75, 76, 79, 107, 124
Orton, William, 63, 107

Pacific Telegraph Act of 1860, 62
Pfisterer, Sophie, 114
Philadelphia-to-Pittsburgh line, 69
Philadelphia-to-Pittsburgh telegraph project, 41–42
Philosophical Italian Society, 98
Pius VIII, Pope, 97
Polk, James K., 39, 101
Pouillat, Alphonse, 23
Pratt, Henry C., 7
Prescott, Thomas, 13
Prime, Samuel I., 36, 110

Reid, James D., 58
Renaissance artists, 4
Reni, Carlo, 4
Rives, William, 83
Rogers, Henry J., 34

Royal E. House, 56

Seebeck, Ernst, 23
Shaffner, Taliaferro Preston, 51–52, 52, 53
Shannon, Claude, 26
Sibley, Hiram, 60, 63
Silliman, Benjamin, 22, 82, 83, 86; study of Weston Meteor, 82
Smith, Francis Ormond Jonathan "Fog", 27, 28–29, 31, 35, 42, 44–45; attempts for telegraph franchises and money troubles, 45–48; scheme for making telegraph attractive, 40
Smith, Reverend E. G., 4
Southwestern Telegraph Co., 56
Speedwell Iron Works, 26
Steinheil, Carl August von, 67
Steinheil, Franz, 31
Sturgeon, Marcel, 67, 78
Swain, thomas, 59

telegraph bill, 31, 32, 34, 110
telegraph system, 22–37; Alexander Bain's telegraph, 59–60; controlled electrical sparks, transmission of, 23; converting sparks into burn coded patterns into paper, 23; electric relay/combined circuit, development of, 25; Henry's work and, 24; House telegraphs, 56–58; inventor's account, 22–23; Morse's functioning machine, 23–24, 25; unveiling of first telegraph, 27; Vail's idea of telegraph transmitter and receiver, 26
Thornton, Edward, 107
trans-Atlantic cable project, 51, 55, 56
Trumbull, Lyman, 8

Vail, Alfred, 26–27, 47, 48, 70–73, 88; *The American Electro-Magnetic Telegraph*, 70; *The Early History of the Electro-Magnetic Telegraph*, 70; final years, 71–72; role in developing telegraph, 70; as Superintendent, 70; writing of Morse code, 70. *See also* Morse Telegraph; telegraph system
Vail, E. Cummings, 73
Vail, George, 72
Vail, William, 72
Vinci, Leonardo da, 1

Wade, Jeptha, 63
Walker, Lucretia, 5, 11, 110
Walker, Thomas R., 50, 62, 79, 81
Washington-Baltimore line, 69
Washington & New Orleans Telegraph Co., 70, 76, 102, 103, 104
Washington-to-Baltimore telegraph project, 35–37, 39; operating budget and revenues, 39–40, 41
Washington-to-New Orleans telegraph project, 41
West, Benjamin, 3
Western Union Telegraph Company, 60–64
Wheatstone, Charles, 67. *See also* Wheatstone-Cooke telegraph
Wheatstone-Cooke telegraph, 29–30, 31, 56; patent, 33
Wilberforce, William, 94
Wood, George, 102, 110
Woodbury, Levi, 27
Wright, Silas, 39

About the Author

George F. Botjer (B.S., Economics, M.A., History, New York University; Ph.D., History, Florida State University) is the author of three previous books: *A Short History of Nationalist China, 1919–1949* (Putnam), *Sideshow War: the Italian Campaign, 1943–1945* (Texas A&M University Press), and *Guide to the American Economy, 1607–2007* (Kindle electronic book). He is professor of history (emeritus) at the University of Tampa, and has also taught at Florida State University.

CPSIA information can be obtained at www.ICGtesting.com
Printed in the USA
BVOW07*1349270415

397725BV00001B/1/P